Marine Science

for Cambridge International AS & A Level

WORKBOOK

Matthew Parkin, Jules Robson & Paul Roobottom

CAMBRIDGE
UNIVERSITY PRESS

University Printing House, Cambridge CB2 8BS, United Kingdom

One Liberty Plaza, 20th Floor, New York, NY 10006, USA

477 Williamstown Road, Port Melbourne, VIC 3207, Australia

314–321, 3rd Floor, Plot 3, Splendor Forum, Jasola District Centre, New Delhi – 110025, India

103 Penang Road, #05-06/07, Visioncrest Commercial, Singapore 238467

Cambridge University Press is part of the University of Cambridge.

It furthers the University's mission by disseminating knowledge in the pursuit of education, learning and research at the highest international levels of excellence.

www.cambridge.org
Information on this title: www.cambridge.org/9781108790499

© Cambridge University Press 2020

This publication is in copyright. Subject to statutory exception and to the provisions of relevant collective licensing agreements, no reproduction of any part may take place without the written permission of Cambridge University Press.

First published in 2020

20 19 18 17 16 15 14 13 12 11 10 9 8 7 6 5 4 3 2

Printed in Mexico by Editorial Impresora Apolo, S.A. de C.V.

A catalogue record for this publication is available from the British Library

ISBN 978-1-108-79049-9 Workbook

Cambridge University Press has no responsibility for the persistence or accuracy of URLs for external or third-party internet websites referred to in this publication, and does not guarantee that any content on such websites is, or will remain, accurate or appropriate. Information regarding prices, travel timetables, and other factual information given in this work is correct at the time of first printing but Cambridge University Press does not guarantee the accuracy of such information thereafter.

....

NOTICE TO TEACHERS IN THE UK
It is illegal to reproduce any part of this work in material form (including photocopying and electronic storage) except under the following circumstances:
(i) where you are abiding by a licence granted to your school or institution by the Copyright Licensing Agency;
(ii) where no such licence exists, or where you wish to exceed the terms of a licence, and you have gained the written permission of Cambridge University Press;
(iii) where you are allowed to reproduce without permission under the provisions of Chapter 3 of the Copyright, Designs and Patents Act 1988, which covers, for example, the reproduction of short passages within certain types of educational anthology and reproduction for the purposes of setting examination questions.

....

This Workbook is designed to support the Coursebook, with specially selected topics where students would benefit from further opportunities to apply skills, such as application, analysis and evaluation in addition to developing knowledge and understanding. (The Workbook does not cover all topics in the Cambridge International AS & A Level Marine Science syllabus (9693)).

Exam-style questions and sample answers have been written by the authors. In examinations, the way marks are awarded may be different. References to assessment and/or assessment preparation are the publisher's interpretation of the syllabus requirements and may not fully reflect the approach of Cambridge Assessment International Education.

Cambridge International copyright material in this publication is reproduced under licence and remains the intellectual property of Cambridge Assessment International Education.

Cambridge International recommends that teachers consider using a range of teaching and learning resources in preparing learners for assessment, based on their own professional judgement of their students' needs.

> Contents

Section 1 — 1

Chapter 1: Water — 2
Exercises — 2
Exam-style questions — 14

Chapter 2: Earth processes — 19
Exercises — 19
Exam-style questions — 26

Chapter 3: Interactions in marine ecosystems — 28
Exercises — 28
Exam-style questions — 38

Chapter 4: Classification and biodiversity — 42
Exercises — 42
Exam-style questions — 53

Chapter 5: Examples of marine ecosystems — 56
Exercises — 56
Exam-style questions — 64

Chapter 6: Physiology of marine organisms — 67
Exercises — 67
Exam-style questions — 74

Chapter 7: Energy — 79
Exercises — 79
Exam-style questions — 85

Chapter 8: Fisheries for the future — 90
Exercises — 90
Exam-style questions — 95

Chapter 9: Human impacts on marine ecosystems — 98
Exercises — 98
Exam-style questions — 109

Section 2   115

Chapter 1: Water   116
Practical 1.1	Properties of water	116
Practical 1.2	pH	119
Practical 1.3	Salinity and temperature gradients	122

Chapter 2: Earth processes   124
Practical 2.1	Investigating the effect of temperature on the solubility of a salt	124
Practical 2.2	Modelling weathering and erosion	126
Practical 2.3	Interpreting tide tables	131

Chapter 3: Interactions in marine ecosystems   133
Practical 3.1	Pyramids of numbers and biomass	133
Practical 3.2	Planning an investigation to estimate the productivity of an aquatic producer	135
Practical 3.3	Investigating the carbon cycle	138

Chapter 4: Classification and biodiversity   143
Practical 4.1	Constructing a dichotomous key	143
Practical 4.2	Using quadrats to estimate abundance of organisms	145
Practical 4.3	Estimating a population size using the mark–release–recapture method	148

Chapter 5: Examples of marine ecosystems   150
Practical 5.1	Drawing an animal found on a sandy shore	150
Practical 5.2	Planning an investigation into the effect of light intensity on coral growth	152
Practical 5.3	Distribution of organisms on a rocky shore	154

Chapter 6: Physiology of marine organisms   159
Practical 6.1	Observing, drawing and comparing the structures of respiratory systems	159
Practical 6.2	Investigating the effect of salinity on brine shrimp	162
Practical 6.3	The effect of salt solution on eggs	163

Chapter 7: Energy   166
Practical 7.1	Identification and separation of photosynthetic pigments using paper chromatography	166
Practical 7.2	Data analysis into limiting factors for photosynthesis	170
Practical 7.3	Gas exchange in an aquatic producer	174

Chapter 8: Fisheries for the future — 177

Practical 8.1   Determining of size of reproductive maturity to inform minimum catch size — 177

Practical 8.2   Effect of temperature on growth of whelk — 180

Practical 8.3   Planning an investigation into the effect of feeding rates on the growth rates of salmon — 183

Chapter 9: Human impacts on marine ecosystems — 185

Practical 9.1   Planning an investigation into marine plastics pollution — 185

Practical 9.2   Modelling the greenhouse effect — 186

Practical 9.3   Monitoring invasive species — 188

Glossary — 193

Acknowledgements — 198

CAMBRIDGE INTERNATIONAL AS & A LEVEL MARINE SCIENCE: WORKBOOK

> How to use this series

This suite of resources supports students and teachers following the Cambridge International AS & A Level Marine Science syllabus (9693). All of the books in the series work together to help students develop the necessary knowledge and scientific skills required for this subject.

The coursebook covers the full Cambridge International AS & A Level Marine Science syllabus, with the chapter structure following the syllabus order. Each chapter includes exercises to develop problem-solving skills; practical activities to help students develop investigative skills; and international case studies and projects to illustrate phenomena in real-world situations. There is a new practical skills chapter that introduces students to experimental planning, presenting data and evaluating experimental methods, with examples and questions.

The teacher's resource supports and unlocks the projects, questions and practical activities in the coursebook, as well as providing detailed lesson ideas and plans. It includes support notes and sample data for the practical activities in the workbook and coursebook. It also contains answers to all questions in the coursebook and workbook.

The workbook contains engaging exercises and exam-style questions to develop scientific skills such as problem-solving, handling and applying information, and mathematical skills for science. It also contains practical activities for each syllabus topic area, to support students' investigative skills, including planning experiments and data exercises.

> How to use this book

Throughout this section, you will notice lots of different features that will help your learning and advance your skills. These are explained below.

Section 1

CHAPTER OUTLINE

These appear at the start of every chapter to introduce the learning aims and help you navigate the content.

Exercises

These help you to practise skills that are important for studying Cambridge International AS & A Level Marine Science.

TIPS

The information in these boxes will help you complete the exercises, and give you support in areas that you might find difficult.

EXAM-STYLE QUESTIONS

Questions at the end of each chapter are more demanding exam-style questions, some of which may require use of knowledge from previous chapters. Answers to these questions can be found in the Teacher's Resource.

KEY WORDS

Key vocabulary is highlighted in the text when it is first introduced. Definitions are then given in the margin, which explain the meanings of these words and phrases.

You will also find definitions of these words in the Glossary at the back of this book.

COMMAND WORDS

Words that might be used in exams are highlighted in the exam-style questions when they are first introduced. In the margin, you will find the definition of these words.

---

* The information in this section is taken from the Cambridge International syllabus for examination from 2022. You should always refer to the appropriate syllabus document for the year of your examination to confirm the details and for more information. The syllabus document is available on the Cambridge International website at www.cambridgeinternational.org.

> How to use this book

Throughout this section, you will notice lots of different features that will help your learning and advance your skills. These are explained below.

Section 2

CHAPTER OUTLINE

These appear at the start of every chapter to help you navigate the content.

Practicals

These help you to develop the practical skills which are essential for studying Cambridge International AS & A Level Marine Science. The investigations contain an introduction which outlines the theory behind the practical work, a list of equipment, important safety advice to ensure you stay safe whilst conducting practical work, a step-by-step method, and finally evaluation questions which help you to interpret your results. Some chapters also contain planning investigations, which allow you to practise planning your own practical work.

REFLECTION

These appear at the end of each practical and contain reflective questions, which allow you to discuss or reflect on the investigation and what you have learnt.

Note: This type of box shows extension content that is not part of the syllabus.

KEY WORDS

Key vocabulary is highlighted in the text when it is first introduced. Definitions are then given in the margin, which explain the meanings of these words and phrases.

You will also find definitions of these words in the Glossary at the back of this book.

TIPS

The information in these boxes will help you complete the investigations, and give you support in areas that you might find difficult.

AS & A Level
Section 1

SECTION OUTLINE

The Cambridge International Marine Science AS and A Level syllabus covers a wide range of topics and require a range of academic skills. This section of the workbook is designed to help you develop your detailed knowledge of the topic areas and the skills that will help you tackle your course and the examinations. Skills covered include: methods of data analysis, investigation planning, mathematical methods and statistical testing. The exercises will guide you in a step-by-step way so that you will feel more confident when handling data and analysing unfamiliar situations. We recommend that you use the workbook alongside the coursebook to help consolidate your knowledge and understanding.

Chapter 1
Water

CHAPTER OUTLINE

The questions in this chapter cover the following topics:
- the composition of seawater
- ionic and covalent bonds
- the properties of water
- factors affecting the density of water
- haloclines and thermoclines.

Exercises

Exercise 1.1 Comparing the ionic composition of seawater by calculating percentages and drawing bar charts

This exercise will help you to compare the **solute** composition of seawater and represent this as a bar chart.

Seawater has many different solutes dissolved in it. The proportions and masses of each of these solutes can differ in different bodies of water. Table 1.1 shows the masses of solutes found in 1 dm^3 of seawater taken from three different bodies of water. Notice these two important features of the data:

- The different seawaters have different overall solute concentrations as the total amount of solute varies.
- The proportions of each individual solute are different.

KEY WORD

solute: a solid that dissolves in a solvent

Solute	Mass of solute dissolved in 1 dm^3 / mg		
	Atlantic Ocean	Mediterranean Sea	Sea off coast of Kuwait
chloride	18 980	21 200	23 000
sodium	10 556	11 800	15 850
sulfate	2 649	2 950	3 200
magnesium	1 262	1 403	1 765
calcium	400	423	500
potassium	380	463	460
bicarbonate	140	143	142
others	98	229	92
Total	34 465	38 611	45 009

Table 1.1: The solute composition of different bodies of seawater.

If you want to compare the proportions of the solutes of different samples of seawater, you could directly compare the masses of the solutes. This is valid if the overall volumes of the samples are the same, but not a fair comparison if they are different. The mass of chloride ions dissolved in one cubic decimetre of seawater will obviously be higher than in one cubic centimetre, simply because there is more seawater. Also, the overall concentrations (salinities) of the seawaters are different and we are trying to compare the proportion of total solutes that each individual solute takes up. A more valid way of comparing the proportions of the solutes would be to calculate the percentage of each solute compared to the total solutes.

To calculate a percentage, divide the number by the total and then multiply by one hundred.

For example, the total mass of solutes dissolved in Atlantic Ocean seawater in Table 1.1 is 34 465 mg.

To calculate the percentage of dissolved solute that consists of chloride ions, divide the mass of chloride by the total mass and then multiply by 100:

$$\text{percentage of chloride ions} = \frac{18\,980}{34\,465} \times 100\% = 55.1\%$$

1 a Calculate the percentages of all the solutes for the three different bodies of water shown in Table 1.1. Copy and complete Table 1.2.

  b Compare and contrast the percentages of the solutes by identifying clear similarities and differences.

Solute	Percentage of each solute / %		
	Atlantic Ocean	Mediterranean Sea	Sea off coast of Kuwait
chloride	55.1		
sodium			
sulfate			
magnesium			
calcium			
potassium			
bicarbonate			
others			

Table 1.2: Percentages of solutes in different bodies of seawater.

Sometimes, it is useful to view data as a chart. Bar charts are used when one variable is categoric (also known as discontinuous) and the other variable is continuous. A **categoric variable** is something that has a particular category, or name (for example, the names of the solutes listed). A **continuous variable** is something that has numbers that can have any number of intermediate values, such as height, weight or percentage.

2 a Follow the steps below to draw a bar chart for the data in Table 1.1. In this example, you will compare the masses of chloride, sodium, sulfate and magnesium in the different bodies of seawater.

KEY WORDS

categoric variable: a variable that is not continuous and has a value that is a name or label such as the colours red, blue and green

continuous variable: one which can take any value (for example, temperature, time, concentration)

Step 1: Decide which way round the axes will go. Although not essential, it is often good practice to place the **independent variable** on the horizontal (x) axis and the **dependent variable** on the vertical (y) axis. The data can be organised on the horizontal axis in different ways, depending on what you are comparing. For the data in Table 1.1, we want to compare the masses of each solute in the different bodies of seawater. We could group the data as similar solutes or as water bodies. In this example, you will group the data as solutes.

Step 2: Label the vertical axis 'mass of solute (mg dm^{-3})' and the horizontal axis 'solute'. Decide on a suitable linear scale for your continuous variable. You should identify the maximum and minimum values and choose a scale that makes the best use of your graph paper, or grid. Ideally the graph should use at least half the grid. Scales should be always be linear, with even increments. For the data in Table 1.1, use increments of 5000 mg dm^{-3}, starting at 0 mg dm^{-3} and ending at 25 000 mg dm^{-3}.

b   Draw the scale for the y-axis.

Step 3: Plot and draw the bars. Decide how many bars you need to fit onto your graph paper. There are 12 bars required for the data in Table 1.1 (three each for chloride, sodium, sulfate and magnesium.) Bars should be drawn with a sharp pencil and a ruler. They should not touch and should have equal gaps between them within the groups of data. In this example, you should group the bars for each body of seawater for each solute. Leave a slightly bigger gap between the three bars for each solute, as shown in Figure 1.1, which shows an example of a bar chart. The groups should be labelled with the correct solute underneath the horizontal axis.

c   Draw your bars on your graph and add labels.

> **KEY WORDS**
>
> **independent variable:** the variable being changed in an experiment
>
> **dependent variable:** the variable being measured in an experiment

> **TIP**
>
> When selecting scales, always use linear scales with even increments. Pick sensible increments that make it easy to plot points.

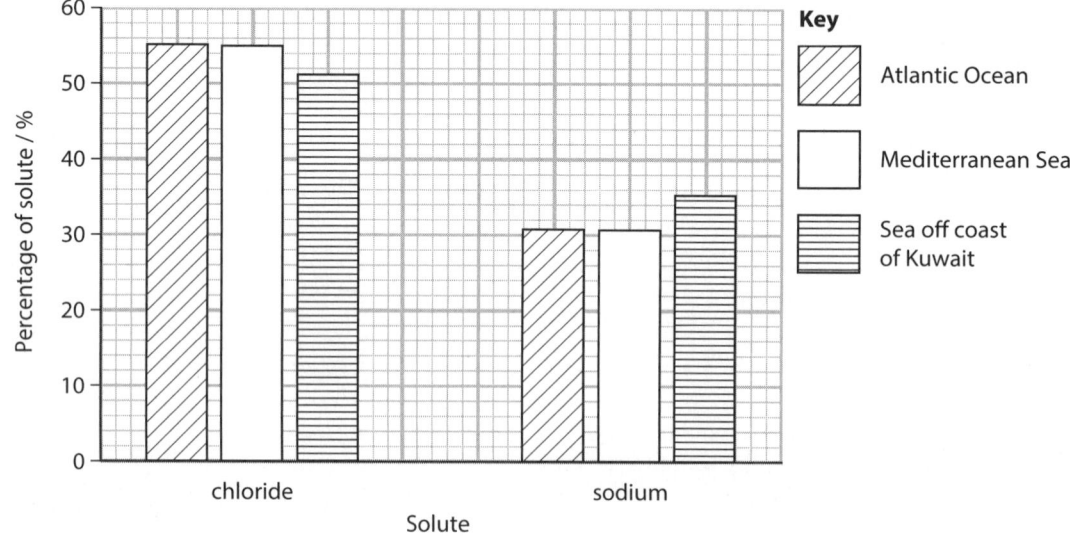

Figure 1.1: Grouping of bars for a bar chart.

Step 4: Decide on a key for the different bars. For this example, choose a colour or pattern for each of the bodies of water, draw out a key on the chart and shade the bars appropriately. Figure 1.1 shows a bar chart of the percentages of sodium and chloride in the different areas. The bars are grouped as the solutes and a key to show each area of water is shown.

- **d** Produce a key and shade or colour the bars.
- **e** Produce a second bar chart. This time, group the bars as the different bodies or seawater so that the horizontal (x) axis is labelled 'body of seawater'.
- **f** Look at your bar charts and the information in Table 1.1. Compare the masses of the solutes from the three bodies of seawater. Give similarities and differences.

3 Figure 1.2 shows the locations of the bodies of seawater. Use your knowledge of factors that affect solute concentrations to suggest reasons for the differences in composition of the water from the three bodies of seawater. Consider both the proportions and the overall concentrations of the seawaters.

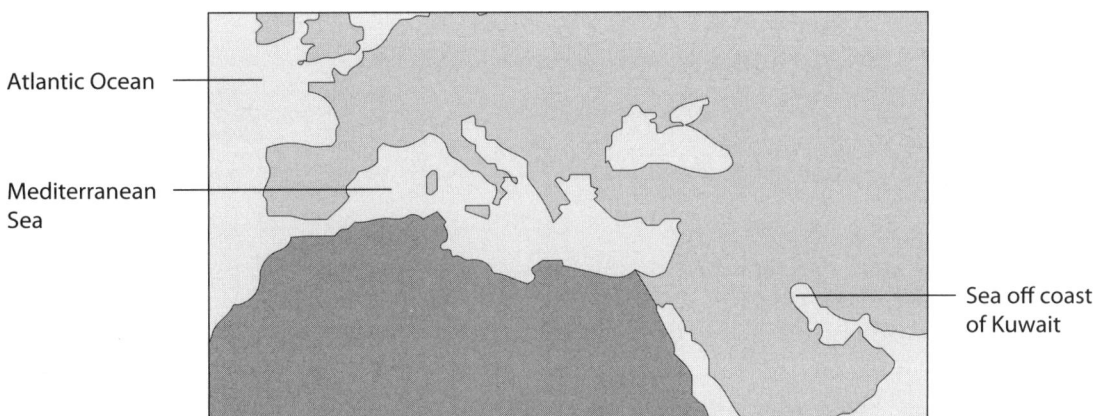

Figure 1.2: Locations of the three bodies of seawater.

Exercise 1.2 Understanding different variables and planning investigations

In this exercise, you will develop your understanding of the different types of variables that are considered when planning scientific investigations.

We consider three main types of variables when planning experimental investigations:

- Independent variable: This is the variable that the experimenter changes. Ideally, you should plan for at least five different values, with even increments between them. You also need a range of values that will produce a valid conclusion.

- Dependent variable: This is the variable that the experimenter measures after changing the independent variable. For reliable experiments, at least three replicates should be obtained.

- **Standardised variables:** These are other variables that could affect the results of the experiment. To generate valid data, these should be kept constant and you should also state how you will do this in your plan.

> **KEY WORD**
>
> **standardised variable:** a variable that is kept constant during an experiment

1   Read the following student practical.

<u>An investigation into the effect of water temperature on the solubility of oxygen.</u>

I will use a measuring cylinder to measure out 150 cm³ of 3% sodium chloride solution. I will place the solution into a 250 cm³ beaker. I will place the beaker into a thermostatically controlled waterbath and set the waterbath to 5°C. I will then bubble oxygen into the water for five minutes (timed with a stopclock). After the five minutes, I will measure the concentration of oxygen in the water with an oxygen meter for 30 seconds. I will repeat this at temperatures of 10°C, 15°C, 20°C, 25°C, 30°C and 35°C. I will repeat the experiment three times to obtain mean values.

a Give the independent variable that the student changed. Describe how the independent variable was changed, and the number and range of different values used.

b Give the dependent variable. Describe how the student measured the dependent variable and how they ensured reliability.

c List the standardised variables. For each standardised variable, state how it was kept constant and why it was kept constant.

2   Use the information in parts **1a**, **b** and **c** to plan an investigation into the effect of salinity on the **solubility** of oxygen.

Exercise 1.3 Atomic structure and chemical bonding

To understand the composition of seawater, you need to have a full understanding of atomic structure and the nature of **ionic bonds** and **covalent bonds**. This exercise will help you to understand atomic structure and how this affects the bonds formed between atoms.

You need to know the locations, charges and masses of the three main subatomic particles (protons, neutrons and electrons.)

1   **a** Copy out Table 1.3 and use your coursebook and your own knowledge to add the:

- relative masses (1 or 0)
- relative charges (+1, –1, 0)
- location of each particle within an atom (nucleus, orbital).

Subatomic particle	Relative mass	Charge	Location within atom
proton			
neutron			
electron			

Table 1.3: Properties of subatomic particles.

The **atomic number** of an element is the number of protons that it has. It is also the number of electrons present, because in any atom, the number of protons is the same as the number of electrons.

The relative **atomic mass** of an element is the number of protons plus the number of neutrons.

KEY WORDS

solubility: the ability of a solute to dissolve within a solvent (such as water)

ionic bond: chemical bond that involves the attraction between two oppositely charged ions

covalent bond: chemical bond that involves the sharing of electron pairs between atoms

atomic number: the number of protons contained in the nucleus of an atom

atomic mass: the mass of an atom that is approximately equal to the number of protons and the number of neutrons added together

For example, the element, sodium (Na) has an atomic number of 11 and a mass number of 23. This means that one atom has:

11 protons

11 electrons

12 neutrons (23 minus 11).

The Periodic Table of the Elements arranges all the elements in order of atomic number and gives the atomic number and relative atomic mass of each element. Figure 1.3 shows a Periodic Table.

Figure 1.3: Periodic Table of the Elements.

b Use Figure 1.3 to determine the numbers of protons, neutrons and electrons in each of the elements that are often found in compounds dissolved in seawater. Copy and complete Table 1.4.

Element	Atomic number	Relative atomic mass	Number of protons	Number of neutrons	Number of electrons
calcium (Ca)					
carbon (C)					
chlorine (Cl)					
hydrogen (H)					
magnesium (Mg)					
nitrogen (N)					
oxygen (O)					

Table 1.4: Properties of common elements present in compounds dissolved in seawater.

> **TIP**
>
> Note that the relative atomic mass of chlorine is 35.5, because some chlorine atoms have a relative atomic mass of 35 and others 36.

Electrons form the chemical bonds between elements. Electrons are arranged in orbitals called shells and the number of electrons that fill up each shell depends on which shell it is. You will only consider the first three shells in this exercise. The innermost shell is full when it contains two electrons. The next two shells are both full when they contain eight electrons each.

We can illustrate the electron shells with Bohr diagrams. The element argon (Ar) has an atomic number of 18. This means that argon has 18 electrons arranged in the shells as 2, 8, 8 (this means there are two electrons in the first shell, eight in the second and eight in the third.) This arrangement of electrons is known as the electron configuration. Figure 1.4 shows a Bohr diagram for argon.

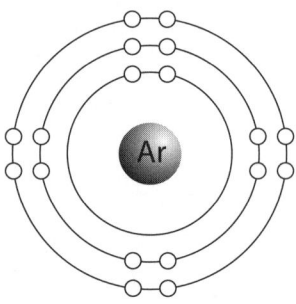

Figure 1.4: Bohr diagram showing 2, 8, 8 electron configuration of argon (Ar). The three electron shells and electrons are shown.

2 Draw Bohr diagrams to show the electron configurations of:
   a hydrogen atom (H)
   b sodium (Na)
   c chlorine (Cl)
   d magnesium (Mg)
   e oxygen (O).

Electron shells are stable when they are full. For the innermost electron shell, this is two electrons and for the next two electron shells, this is eight electrons. During chemical reactions, atoms lose or gain electrons so that the outermost shell is full. If an atom has only one or two electrons in its outermost shell, it will often lose them to become more stable. If an atom has six or seven electrons in its outermost shell, it will often gain electrons from another element. Losing or gaining electrons results in the production of a charged particle known as an ion.

For example, the element lithium (Li) has an atomic number of 3. This means that a lithium atom has three protons, each with a positive charge, and three electrons, each with a negative charge. The overall charge is zero because the three positives cancel out the three negatives. Lithium has a single electron in its outer shell, so to become more stable it can lose this electron. This means that it still has three positively charge protons but now has only two negatively charge electrons, for an overall charge of +1. This is a lithium ion, which we write as Li^+. What does lithium give this electron to? Atoms with six or seven electrons in their outer shell will accept electrons to fill up the

> **TIP**
>
> Electrons fill up from the innermost shell first. Draw circles for each shell. Add electrons to the inner shell first.

shell. For example, fluorine has an electron configuration of 2, 7 and so can take an electron from a lithium atom to produce a fluoride ion (F⁻) with an overall charge of −1. This chemical reaction produces an ionic compound called lithium fluoride, which is composed of Li⁺ and F⁻ ions. Figure 1.5 shows the reaction between lithium and fluorine.

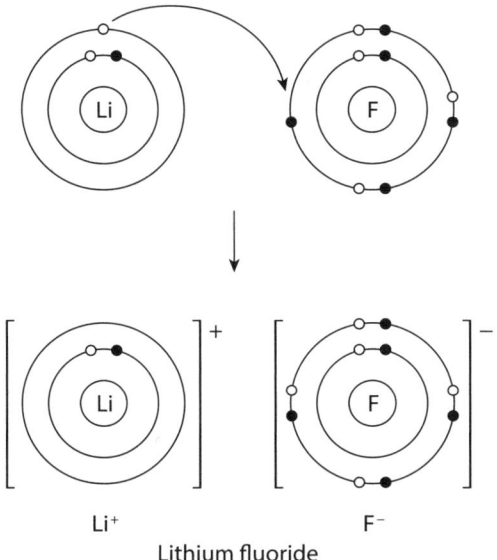

Li⁺   F⁻
Lithium fluoride

Figure 1.5: Reaction between lithium and fluorine.

3 Draw diagrams similar to Figure 1.5 to show the reactions between:
   a   sodium and chlorine
   b   lithium and chlorine
   c   magnesium and chlorine (with two chlorine atoms and one magnesium atom).

In ionic compounds, one ion is negative and the other positive. The two ions attract each other and, when in a solid crystal, they arrange themselves so that positive and negative ions are touching. These are ionic bonds.

   d   Use your knowledge to draw out the crystal structure of sodium chloride.

Sometimes, instead of losing or gaining electrons, atoms can share electrons with other atoms. This sharing of electrons forms covalent bonds that join the atoms in a chemical compound.

Fluorine gas molecules exist as two fluorine atoms bonded together so that their outer shells share two electrons. This is shown in Figure 1.6. The electron shells of each atom overlap and so the two atoms are strongly bound together with a covalent bond. Both atoms have eight electrons in their outer shells as they each share one electron.

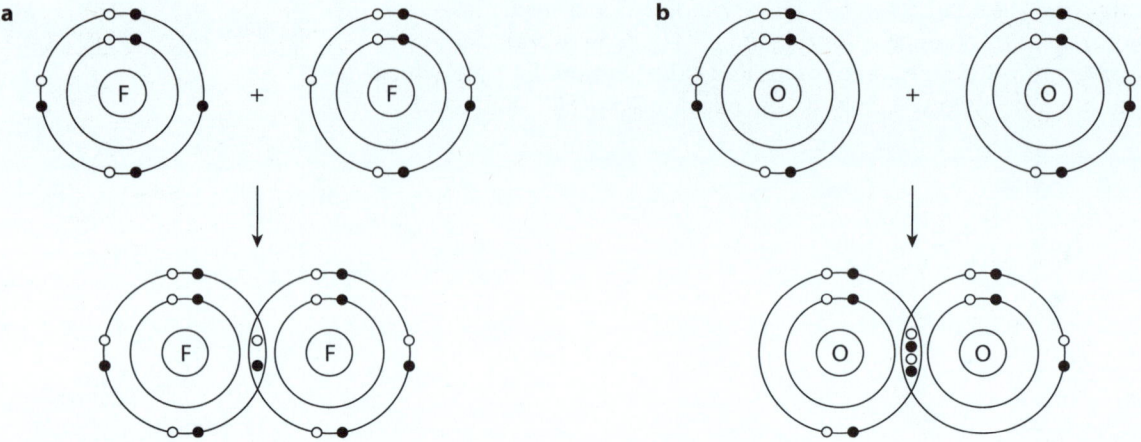

Figure 1.6: Covalent bonds in (a) fluorine molecules (b) oxygen molecules.

Some substances, such as oxygen atoms, can form double bonds where two electrons from each atom are shared (shown in Figure 1.6).

4 Draw diagrams similar to Figure 1.6 to show the bonding in:
   a chlorine (Cl_2)
   b carbon dioxide (CO_2)
   c water (H_2O)
   d sulfur dioxide (SO_2).

When bonds form between oxygen and hydrogen to make water, the oxygen and hydrogen do not share the electrons equally. The electrons are drawn closer to the oxygen atom, so that the oxygen atom develops a slight negative charge and the two hydrogen atoms develop a slight positive charge. This means that water molecules are polar molecules, which are molecules with both positive and negative charges. Because the charges on the water molecule are not whole positive and negative charges, but are small charges, we label them δ+ and δ−. This polar nature of water gives it unusual properties that make it important for life.

5 a Copy out Figure 1.7 and label the oxygen and hydrogen atoms, and the areas with charges of δ+ and δ-.

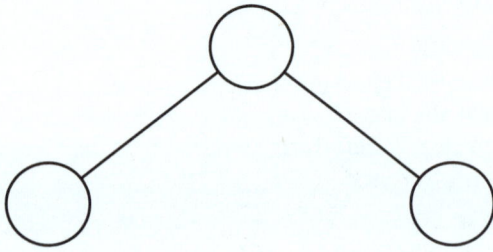

Figure 1.7: The structure of a water molecule.

b Because water molecules have positively and negatively charged areas, water molecules attach to each other by forming **hydrogen bonds**. Draw a diagram with six molecules of water to show how hydrogen bonds form between water molecules.

c Draw a diagram to show how water molecules would associate around a sodium ion and a chloride ion. Label the charges on each ion or atom.

d Explain how the polar nature of water affects the specific heat capacity, **density** and solvent nature of water.

Exercise 1.4 Analysing data about water density, temperature and salinity

You need to be able to look at data patterns and give accurate descriptions and explanations of any patterns present. This exercise will help develop your data analysis skills.

Figure 1.8 shows the effect of temperature on the density of fresh water. If you are asked to describe the effect of one variable on another, follow these steps:

- Describe the general patterns giving clear indications of directions (for example, 'As *x* increases, *y* increases.'). Use the labels on the axes to clarify your description.
- Look for positive and negative **correlations**. In a positive correlation, an increase in one variable is linked to an increase in another variable. In a negative correlation, an increase in one variable is linked to a decrease in another variable. A correlation means that there may be a link, but the link is not necessarily causal.
- Look for areas where there is a change in pattern (or turning points) such as an increase, decrease or levelling off.

1 Figure 1.8 shows the effect of temperature on the density of fresh water. Describe how temperature affects the density of water.

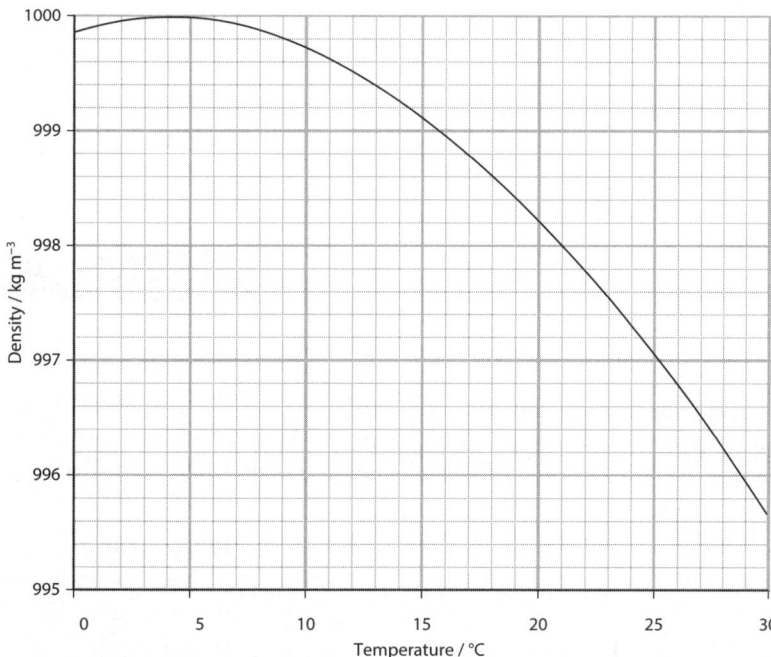

Figure 1.8: The effect of temperature on the density of fresh water.

> **KEY WORDS**
>
> **hydrogen bond:** a weak bond between two molecules due to the electrostatic attraction between a hydrogen atom in one molecule and an atom of oxygen, nitrogen or fluorine in the other molecule
>
> **density:** a measure of the mass of a defined volume of water
>
> **correlation:** the tendency of two variables to change together in either the same or opposite directions

2 Look at the graphs in Figure 1.9. They are an unusual presentation because the *y*-axis seems to be upside down. The zero value is at the top because it illustrates the depth from the surface of the water. The graphs show how temperature, salinity and density change with depth of water. Answer the following questions in detail.

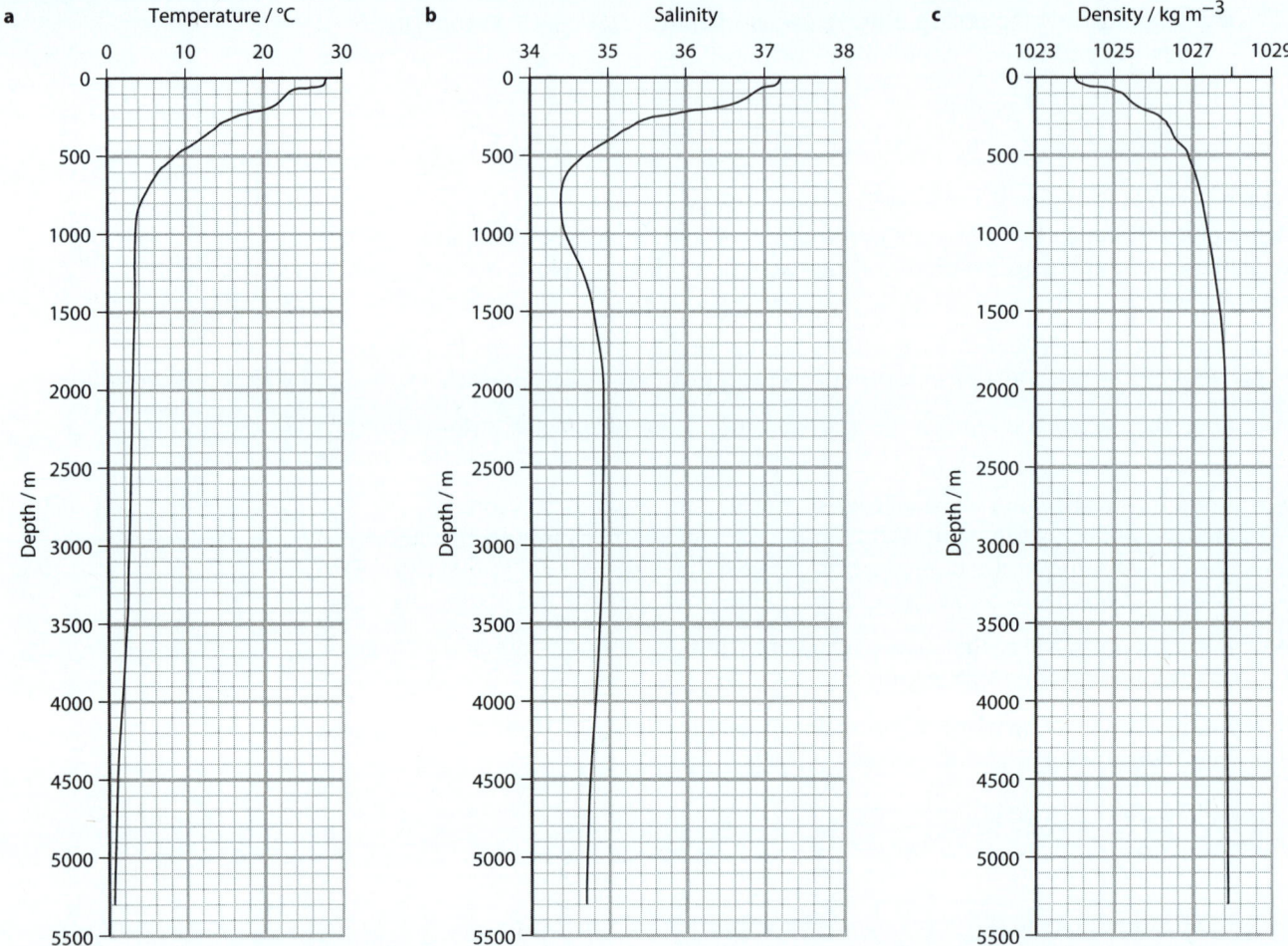

Figure 1.9: Graphs to show (a) temperature, (b) salinity and (c) density at different depths of water

a Describe how temperature changes with depth.
b Describe how salinity changes with depth.
c Describe how density changes with depth.
d Explain which variables show positive and negative correlations.
e State which of the graphs shows a **halocline** and which shows a **thermocline**.

KEY WORDS

halocline: a layer of water below the mixed surface layer where a rapid change in salinity can be measured as depth increases

thermocline: a layer between two layers of water with different temperatures

You need to be able to extract data from graphs and charts that may be presented in varied styles. Figure 1.10 shows the effect of temperature and salinity on the density of water.

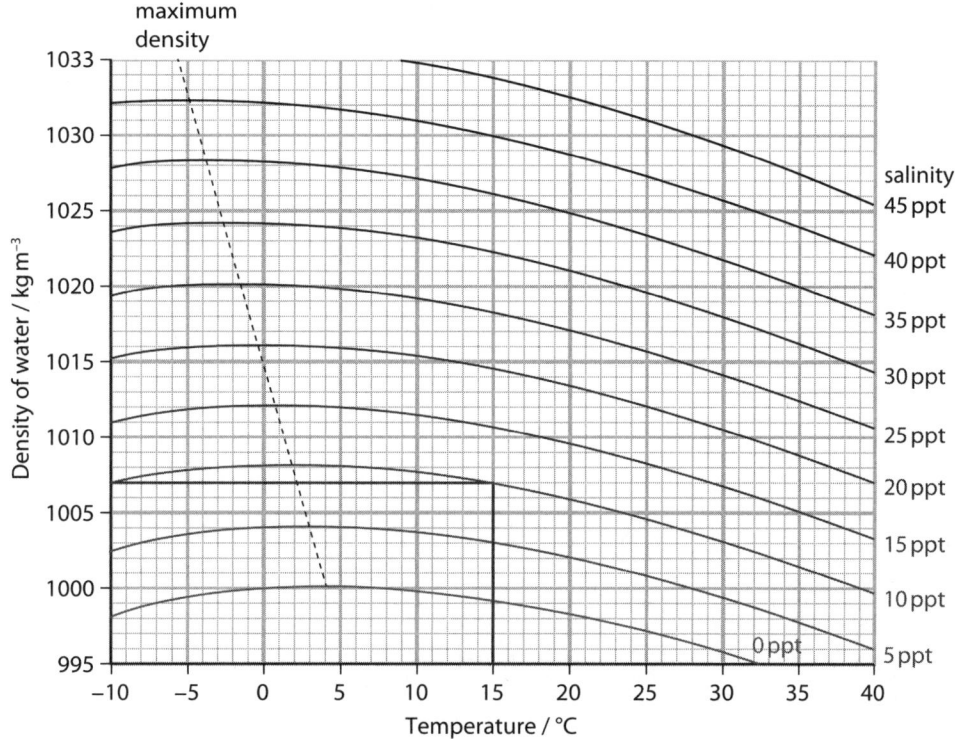

Figure 1.10: The effect of salinity and temperature on the density of water.

You can use the graph to determine the density of water of different salinities at different temperatures. For example, to determine the density of water with a salinity of 15ppt at 15 °C:

- Use a ruler to draw a line up to the 15 ppt curve from 15 °C.
- Draw a line from this point to the vertical axis.
- Read across to the vertical axis to see that the density would be 1007 kg m^{-3}.

3 a Determine the density of water of salinity, 25 ppt, at 20 °C.
  b Determine the density of water of salinity, 35 ppt, at a temperature of 30 °C.
  c Determine the temperature at which water of salinity 20 ppt has a density of 1015 kg m^{-3}.

4 Use the equation for density and Figure 1.10 to determine the mass of 0.5 dm³ of water with salinity of 20 ppt at a temperature of 20 °C.

$$\text{density} = \frac{\text{mass}}{\text{volume}}$$

5 Use Figure 1.10 to describe how the maximum density of water at different temperatures is affected by increasing salinity.

EXAM-STYLE QUESTIONS

1 Figure 1.11 shows the location of an ocean lagoon.

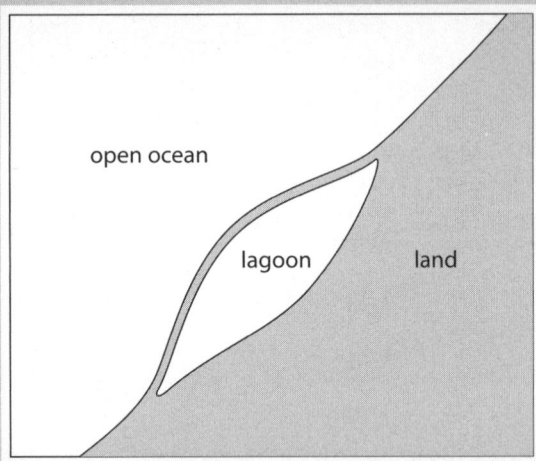

Figure 1.11: Location of an ocean lagoon

The solute concentrations of the lagoon and the open ocean were measured. The results are shown in Table 1.5.

Solute	Solute concentrations / g dm^{-3}	
	Lagoon	Open ocean
sodium	17	12
chloride	25	19
sulfate	5	3

Table 1.5: Solute concentrations in lagoon and open ocean

a Use your knowledge of water and ionic compounds to **explain** how sodium chloride dissolves in water. [4]

b Draw a bar chart to **compare** the solute concentrations of the lagoon and open ocean. [5]

A scientist determined the mean temperature, mass of dissolved oxygen and wave speed of the lagoon and open ocean at 3.00 p.m. each day for one week. The results are shown in Table 1.6.

Area of water	Mean temperature at 3.00 p.m. / °C	Mean concentration of oxygen / mg dm^{-3}	Mean wave speed / m m^{-1}
lagoon	25	6	
open ocean	18	10	24.3

Table 1.6: Mean temperature, oxygen concentration and wave speed of lagoon and open ocean

COMMAND WORDS

explain: set out purposes or reasons / make the relationships between things clear / say why and/or how and support with relevant evidence

compare: identify / comment on similarities and/or differences

1 Water

CONTINUED

The wave speed measurements for the lagoon were:

$2.50\,m\,s^{-1}$, $2.70\,m\,s^{-1}$, $3.90\,m\,s^{-1}$, $1.10\,m\,s^{-1}$, $2.60\,m\,s^{-1}$, $3.80\,m\,s^{-1}$, $3.60\,m\,s^{-1}$

c i **Calculate** the mean wave speed for the lagoon. Give your answer to three significant figures and write it in Table 1.6. [2]

   ii Use the information in Figure 1.11, Table 1.5 and Table 1.6 to **suggest** explanations for the different oxygen concentrations in the lagoon and open ocean. [5]

[Total: 16]

2 Figure 1.12 shows the effects of salinity and temperature on the freezing point and density of water.

The temperature at which liquid water has its maximum density is also shown. In areas above this line, the density of water decreases as temperature increases. In areas below this line, where water is a liquid, the density of liquid water decreases as temperature decreases.

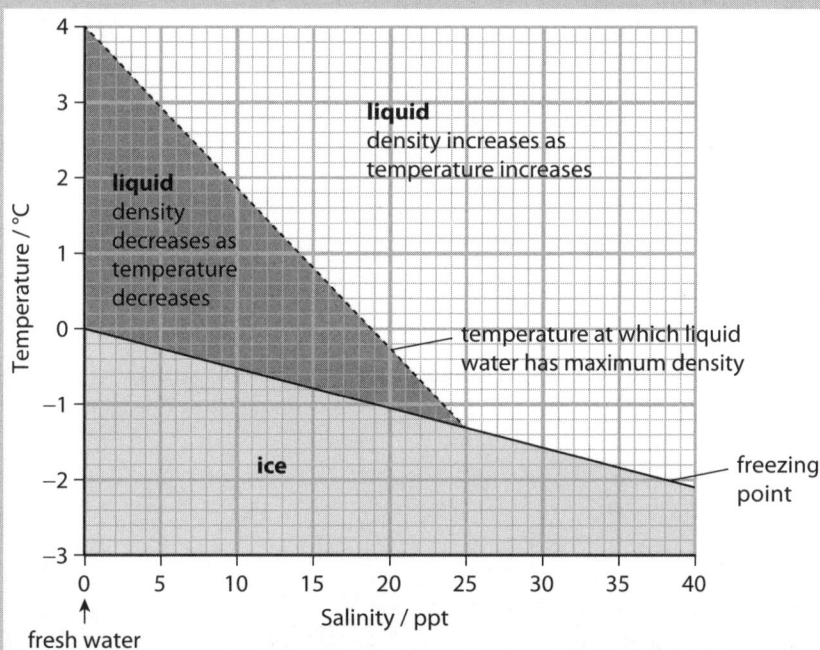

Figure 1.12: Effects of salinity and temperature on the freezing point and density of water

COMMAND WORDS

calculate: work out from given facts, figures or information

suggest: apply knowledge and understanding to situations where there are a range of valid responses in order to make proposals / put forward suggestions

CONTINUED

a i Use Figure 1.12 to **identify** the freezing point of water with a salinity of 35 ppt. [1]

ii **Describe** the effect of increasing salinity on the freezing point of water. [1]

iii **Give** the temperatures at which liquid freshwater (0 ppt salinity) and liquid seawater of 20 ppt salinity, have their maximum densities. [2]

iv Use Figure 1.12 to explain why seawater with a salinity of 35 ppt sinks as it cools but fresh water rises as it cools from 4 °C to freezing point. [2]

b Explain the importance of floating sea ice for marine organisms. [3]

c Describe an investigation into the effect of sodium chloride on the freezing point of water. Include a results table which you could use to record your data. [6]

[Total: 15]

COMMAND WORDS

identify: name / select / recognise

describe: state the points of a topic / give characteristics and main features

give: produce an answer from a given source or recall / memory

3 Table 1.7 shows the concentrations of oxygen at different water depths in a region of the Pacific Ocean.

Depth / m	Concentration of oxygen / mg dm^{-3}
0	6.2
250	5.8
500	3.5
750	2.0
1000	1.1
1500	1.5
2000	1.7
3000	1.9
4000	2.0

Table 1.7: Change in oxygen concentration with increasing ocean depth

a i Describe the change in concentration of oxygen with increasing depth. [3]

ii Calculate the mean change in concentration of oxygen per metre between 0 metres and 1000 metres. [2]

iii Suggest reasons for the changes in oxygen concentration between 0 metres and 1000 metres. [4]

iv Suggest reasons for the changes in oxygen concentration between 1000 metres and 4000 metres. [3]

CONTINUED

b Describe how the effect of depth on the acidity of the water could be tested. [2]

[Total: 14]

4 a Magnesium ions (Mg^{2+}) are a solute of seawater.

The atomic number of magnesium is 12.

The relative atomic mass of magnesium is 24.

i **State** the number of protons, neutrons and electrons in a magnesium ion (Mg^{2+}). [3]

ii Copy and complete Table 1.8. Place a tick in the correct column of Table 1.8 to identify each of these substances found in seawater as a covalent molecule or an ionic substance. [2]

Substance	Covalent molecule	Ionic substance
calcium carbonate		
carbon dioxide		
magnesium sulfate		
oxygen		
sulfur dioxide		
water		

Table 1.8: Substances found in seawater

iii Explain how the structure of water enables it to form hydrogen bonds. [3]

iv Explain how the structure of water affects its ability to act as a solvent. [3]

> **COMMAND WORD**
>
> **state:** express in clear terms

CONTINUED

b Figure 1.13 shows the changes in temperature with depth of an area of temperate ocean and an area of tropical ocean in both winter and summer.

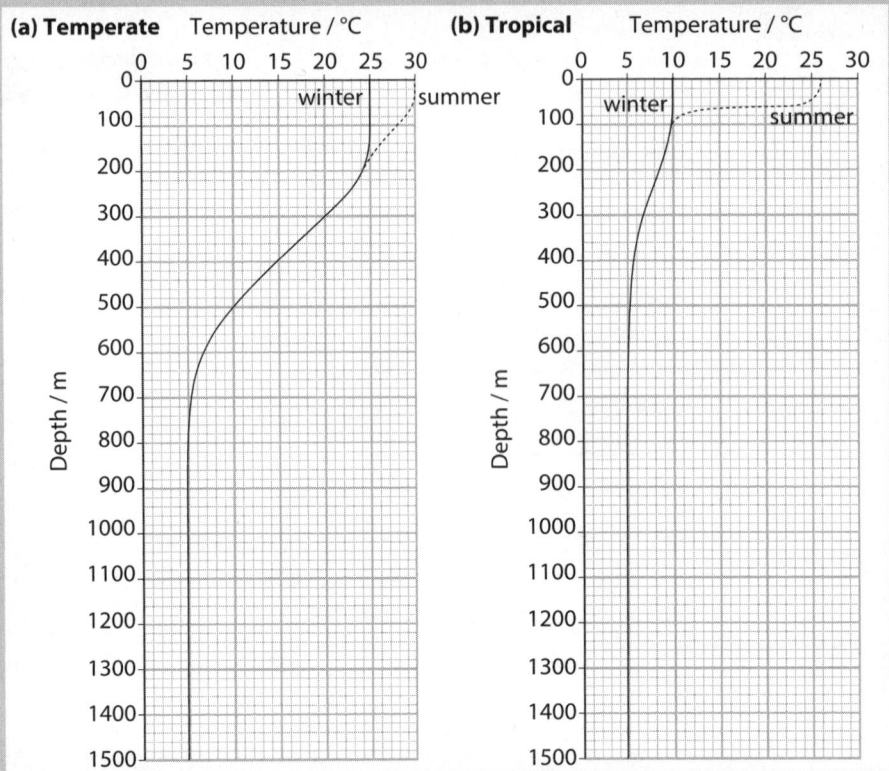

Figure 1.13: Changes in temperature with depth of a temperature and a tropical ocean in both winter and summer

i Estimate the depth limits of the thermocline in summer for the temperate ocean shown in Figure 1.13. [1]

ii **Comment** on the water temperatures in summer and winter for the tropical and temperate oceans shown in Figure 1.13. [4]

iii Explain how thermoclines are produced. [3]

[Total: 19]

COMMAND WORD

comment: give an informed opinion

Chapter 2
Earth processes

CHAPTER OUTLINE

The questions in this chapter cover the following topics:

- the structure of the Earth and how the three types of plate boundary produce tectonic features
- the use of equations, SI units and standard form to answer tectonic processes calculations
- the differing types of weathering and erosion
- how weathering, erosion and sedimentation help shape the formation of rocky, sandy and muddy shores
- the plotting of line graphs and the types of correlation relationships
- the use of tide charts to answer questions on tides and ocean currents
- the principles and evidence for the theory of plate tectonics
- the littoral zone and the formation of deltas and estuaries
- how the alignment of the Earth, moon and sun affects tidal range and produces spring and neap tides
- the causes of the global ocean conveyor belt
- the causes and importance of the El Niño Southern Oscillation.

Exercises

Exercise 2.1 Tectonic processes

This exercise will help describe the structure of the Earth and how the three types of plate boundary produce tectonic features. You will also gain confidence in using mathematical formulae to answer tectonic calculations.

1 **Divergent boundaries** are areas where two tectonic plates are moving away from each other, which may allow molten magma (lava) from the mantle to push through an opening in the crust. The magma spreads out and solidifies in the cold ocean waters to create **mid-ocean ridges**. As more magma is released, the new seafloor is slowly pushed away from the ridge.

   The **spreading rate** of the new seafloor on one side of a mid-ocean ridge is called a half spreading rate. Multiplying it by 2 allows us to calculate the full spreading rate for both sides of the ridge. The rate at which the new seafloor moves away from the mid-ocean ridge can be calculated using the following equation.

   $$\text{spreading rate} = \frac{\text{distance } (d)}{\text{time } (t)}$$

   a **i** Rewrite this formula to calculate distance.

   ii Rewrite this formula to calculate time.

KEY WORDS

divergent boundary: where two tectonic plates are moving away from each other

mid-ocean ridge: a mountain range with a central valley on an ocean floor at the boundary between two diverging tectonic plates, where new crust forms from upwelling magma

spreading rate: the rate at which the new seafloor moves apart at a mid-ocean ridge

b The spreading rate for the seafloor around the Mid-Atlantic Ridge is approximately 2 cm year^{-1}. Calculate in km how far the seafloor will move apart in 10 million years. State your answer in standard form. Show your working.

Step 1: Calculate the distance in cm.

Step 2: Convert the distance from cm to km.

Step 3: State your answer in standard form.

c Calculate the half spreading rate in mm year^{-1} for the Pacific Ocean if the spreading rate of the new seafloor moving apart at a mid-ocean ridge is 1300 km in 10 million years. Show your working.

Step 1: Convert the distance from km to mm.

Step 2: Convert the spreading rate into mm year^{-1}.

Step 3: Calculate the half spreading rate in mm year^{-1}.

d Figure 2.1 shows the seabed spreading of the Mid-Atlantic Ridge. Calculate the half spreading rate.

> **TIP**
>
> Remember to always show your working in full; that is, include any equation used as well as the data you use in your calculation.
>
> Make sure that you have the correct units for each answer; for example, rate must be mm year^{-1} rather than simply mm.

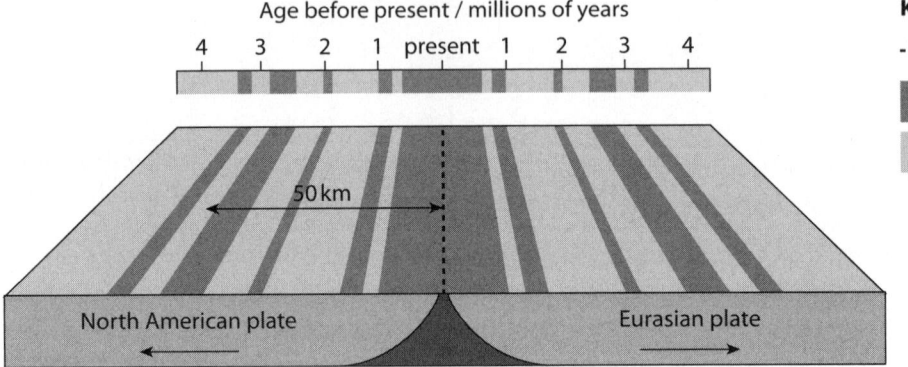

Figure 2.1: Seabed spreading at Mid-Atlantic Ridge.

2 a Sketch and label a diagram of the internal structure of the Earth.

   b Use the information in Table 2.1 and the equation below to calculate the volume of the mantle.

$$V = \frac{4}{3}\pi r^3$$

r = radius
d = diameter

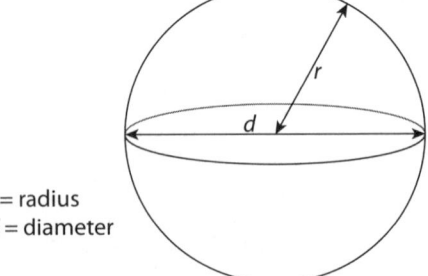

> **TIP**
>
> When drawing, make sure that the scale of each internal structure is appropriate.

Earth's interior	Maximum distance from Earth's surface / km
crust	50
mantle	2900
centre of the Earth	6378

Table 2.1: Distance of Earth's interior layers from Earth's surface.

Step 1: Calculate in km the radius from the centre of the Earth to the inner edge of the mantle (i.e. the core-mantle boundary). Use this radius to calculate the volume of the core.

Step 2: Calculate in km the radius from the centre of the Earth to the far edge of the mantle (i.e. the crust-mantle boundary). Use this radius to calculate the volume of the core and mantle combined.

Step 3: Subtract the answer to Step 1 from the answer to Step 2. This will give the volume of the mantle only.

Step 4: Calculate in km^3 the volume of the centre of the Earth to the crust–mantle boundary.

Step 5: Subtract the answer to Step 4 from the answer to Step 3.

3   Draw a table to describe the differences between **oceanic crust** and **continental crust** with respect to the following characteristics:
- location
- density
- geology
- thickness.

4   Copy and complete Table 2.2 by ticking the type of plate boundary that causes each tectonic feature.

Tectonic feature	Plate boundaries		
	Convergent	Divergent	Transform
ocean trench			
mid-ocean ridge			
hydrothermal vent			
abyssal plains			
underwater volcanoes			
underwater earthquakes			
tsunamis			

Table 2.2: Tectonic features and types of plate boundary.

5   Explain how and where the following tectonic features are formed:
   a   underwater volcanoes
   b   underwater earthquakes
   c   abyssal plains
   d   hydrothermal vents.

KEY WORDS

oceanic crust: The dense, basaltic layer of crust that makes up the bottom of the ocean basins

continental crust: the thicker, less dense crust that makes up the foundation of the continents

TIP

Remember that some tectonic features may be caused by more than one plate boundary.

Exercise 2.2 Erosion and sedimentation

This exercise will explore how weathering, erosion and sedimentation help mould the range of different coastal habitats. It will also build your confidence in performing calculations and plotting line graphs.

1   Describe the terms *weathering*, *erosion* and *sedimentation*.
2   a   Describe the three types of weathering: physical, chemical and organic.
    b   Explain three causes of physical weathering.
    c   Explain three causes of chemical weathering.
    d   Explain three causes of organic (biological) weathering.
3   Explain how each of the following are agents of erosion:
    a   wind
    b   water
    c   ice
    d   gravity.
4   Describe how weathering, erosion and sedimentation give rise to the morphology of:
    a   rocky shores
    b   sandy shores
    c   muddy shores.
5   a   **Sedimentation rates** can be calculated using the following formula:

$$\text{sedimentation rate} = \frac{\text{depth}}{\text{time}}$$

   i   Rearrange the formula to calculate sedimentation depth.
   ii Rearrange the formula to calculate time.

   b   If sedimentation rate is 20 mm/year, what is the depth in km of sediment on the seafloor that is 3 million years old? Show your working.
   Step 1: Calculate the sedimentation depth in mm.
   Step 2: Convert the sedimentation depth to km.

   c   Table 2.3 shows the erosion of a sandy beach over time. This can be measured by the decrease in sand over time.

Year	Depth of sand / cm
1910	70
1930	62
1950	53
1970	46
1990	22
2010	10

Table 2.3: Erosion of a sandy beach over time.

   i   Plot the data for depth of sand vs. year as a line graph.
   ii Use the graph to predict the depth of sand in 1980.

> **TIP**
> Make sure that you focus on describing and explaining the differences between the three types of shores, rather than simply listing factors that are common to all littoral environments.

> **KEY WORD**
> **sedimentation rate:** the rate at which sediment is deposited on the sea floor

6 a Figure 2.2 shows three types of relationships shown by graphs. Describe the type of relationships shown in graphs A, B and C.

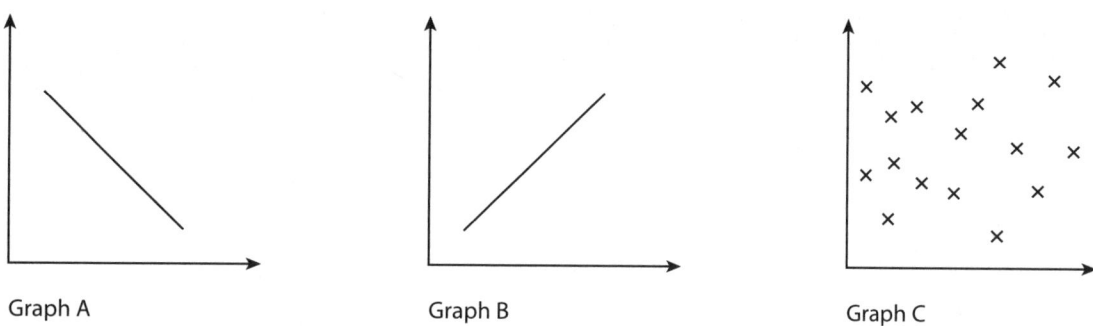

Graph A   Graph B   Graph C

Figure 2.2: Types of graphs.

b State which graph best represents how each of the following factors affects the rate of erosion and sedimentation:
  i particle size
  ii water speed
  iii light intensity.

Exercise 2.3: Tides and ocean currents

This exercise will increase your understanding of tides and ocean currents. You will test your knowledge of definitions before proceeding to gain practice in drawing of line graphs, as well as describing the types of relationship shown by line graphs.

1 Copy Table 2.4 and state whether each of the definitions is True (T) or False (F).

Key word	T / F
Coriolis effect: a force that results from the Earth's rotation that causes objects or particles in motion to deflect to the left in the Northern Hemisphere	
El Niño: a warm current that develops off the coast of Ecuador around December	
neap tide: a tide that occurs when the moon and sun are aligned with each other, causing the smallest tidal range	
littoral zone: the intertidal benthic zone between the highest and lowest spring tide water marks on a shoreline	

Table 2.4: Key word definitions.

2   Ocean currents and tides are both affected by environmental factors such as winds. Table 2.5 shows how wind speed varies with wave height.

Wind speed / km hour^{-1}	Wave height / m
20	0.3
30	0.9
40	1.8
50	3.2
60	5.1
70	7.4
80	10.3
90	13.9

Table 2.5: How wind speed varies with wave height.

   a   Plot the data for wind speed vs. wave height as a line graph.

   Step 1: Choose a scale so that, when the data is plotted, the line will cover at least half the area of the graph paper. The wind speed does not need to start at 0 km hour^{-1}. Choose a simple linear scale, such as setting each small square as equal to 1, 2 or 10 units in the data. Do not use odd numbers.

   Step 2: Choose the axis correctly. Remember to write the names of the x and y axes together with their units. Copy the units carefully from the column heading in the table of data.

   Step 3: Be precise in your plotting; plot the points in the correct location using a sharp pencil. Draw the points lightly so that you can rub them out if you need to. Make them darker when you are sure that they are right. Use a cross or a dot in a circle for your plot points. The plot points should be no larger than half of one of the small squares on the graph paper.

   Step 4: Draw a smooth line through the points using a pencil. Do not extend your line beyond the plotted points; you do not have data for a wind speed of 0 km hour^{-1}, so do not join your line to the origin.

   b   Name and describe the relationship of this graph.

> **TIP**
>
> The x-axis is the independent variable (the factor being changed) whereas the y-axis is the dependent variable (the factor being measured).

c The Douglas Sea Scale shown in Table 2.6 allows us to describe the state of the sea. Use your graph to predict the state of the sea if the wind speed is 58 km hour^{-1}.

Douglas Sea Scale degree	Height / m	Description
0	no wave	calm (glassy)
1	0 – 0.1	calm (rippled)
2	0.1 – 0.5	smooth
3	0.5 – 1.25	slight
4	1.25 – 2.5	moderate
5	2.5 – 4	rough
6	4 – 6	very rough
7	6 – 9	high
8	9 – 14	very high
9	14+	phenomenal

Table 2.6: The Douglas Sea Scale.

Step 1: Make sure to start from the correct axis for wind speed and to work out the scale correctly.

Step 2: Show your working by using a ruler to draw a line vertically up from the correct point on the x-axis until it meets your graph line. Then draw a line horizontally until it joins the y-axis.

Step 3: Carefully read the value for wave height and ensure that your answer includes the correct units.

Step 4: Use the table to interpret the Douglas Sea Scale degree and description for the calculated wave height.

3 a Define tidal range.
  b Explain how each of the following factors can influence the tidal range:
    i depth of water
    ii shape of the coastline
    iii weather.
  c Explain how the alignment of the sun and moon affect the tidal range.
4 Describe the causes of the **global ocean conveyor belt**.

> **KEY WORD**
>
> **global ocean conveyor belt:** constantly moving systems of deep-ocean water driven by thermohaline circulation

EXAM-STYLE QUESTIONS

1   More than 300 million years ago, all the continents on Earth were joined as a single landmass, known as Pangea. Pangea then split into a number of supercontinents including Gondwanaland, which is shown in Figure 2.3.

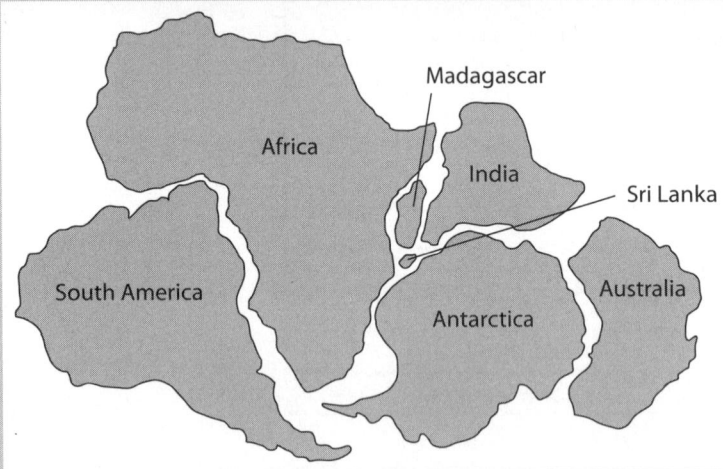

Figure 2.3: A map of Gondwanaland

The landmasses that we recognise today as Africa, South America, Australia, Antarctica, India, Madagascar and Sri Lanka emerged from Gondwanaland.

   a   **Outline** the theory of plate tectonics. [4]
   b   Suggest *three* pieces of evidence in the geology of Gondwanaland that could support the theory that the landmasses were joined together. [3]

   [Total: 7]

2   a   State the meaning of the term *littoral zone*. [1]
   b   **Describe** the difference between estuaries and deltas. [2]
   c   **Describe** how weathering, erosion and sedimentation give rise to the morphology of estuaries and deltas. [5]

   [Total: 8]

COMMAND WORDS

outline: set out main points

describe: state the points of a topic / give characteristics and main features

CONTINUED

3 Figure 2.4 shows part of a tide table with the heights of tides in metres.
   a Calculate the average morning tidal range in this tide table. Show your working. [2]
   b State, giving evidence from the tide chart in Figure 2.4, on which day the neap tide occurs. [2]
   c State, giving evidence from the tide chart in Figure 2.4, on which day the spring tide occurs. [2]
   d Describe the causes of neap and spring tides. [2]

   [Total: 8]

4 The 'normal' direction of wind and ocean currents in the South Pacific is anticlockwise. This spiral pattern is caused by the Coriolis effect. In an El Niño event, the air pressure in the Eastern Pacific drops, causing a reduction in the spiralling ocean waters and winds in the Eastern Pacific. This causes disruption to the oceanic and atmospheric circulation in countries around the Pacific.
   a State what causes the Coriolis effect. [1]
   b Explain what effect El Niño would have on the sardine harvest for Peruvian fishermen in the Eastern Pacific. [4]
   c Explain what effect El Niño would have on the temperature of the New Zealand coast in the Southern Pacific. [2]
   d New Zealand is located at the convergence of subtropical and Antarctic waters. Suggest the effect this has had on biodiversity in its marine waters. [3]
   e The Great Barrier Reef is located on the eastern coast of Australia. In non-El Niño years, the East Australian current brings warm water down the east coast of Australia. **Predict** what effect this would have on the distribution of coral reefs in the north-east of Australia compared with the north-west coastal waters. [1]

   [Total: 11]

Date	Time	Tide height
20 SA ◐	0434	1.7
	1053	4.4
	1704	2.0
	2313	4.7
21 SU	0555	1.8
	1213	4.4
	1830	1.9
22 M	0036	4.7
	0714	1.6
	1331	4.6
	1944	1.7
23 TU	0152	4.8
	0820	1.3
	1434	4.8
	2046	1.4
24 W	0254	5.0
	0918	1.1
	1529	5.0
	2143	1.2
25 TH	0350	5.2
	1011	0.9
	1620	5.2
	2235	1.0
26 F	0442	5.2
	1101	0.8
	1707	5.3
	2323	0.8
27 SA ●	0530	5.3
	1147	0.8
	1747	5.4

Figure 2.4

COMMAND WORD

predict: suggest what may happen based on available information

Chapter 3
Interactions in marine ecosystems

CHAPTER OUTLINE

The questions in this chapter cover the following topics:

- examples of different types of marine symbiotic relationships including parasitism, commensalism and mutualism
- how photosynthetic and chemosynthetic producers make organic energy available to consumers
- feeding relationships terms: trophic level, consumer, producer, herbivore, carnivore, omnivore, decomposer, predator, prey
- calculating primary productivity (GPP and NPP) and the energy losses along food chains (TLTE)
- drawing and interpreting feeding relationships in an ecosystem as food chains and food webs
- drawing, describing and interpreting pyramids of energy, numbers and biomass
- the different types of marine nutrients (gases, inorganic ions and organic compounds)
- how a lack of some elements that are essential to marine life can be a limiting factor to productivity
- the elements that make up carbohydrates, lipids and proteins
- how large molecules are made up of smaller molecules
- the processes by which the reservoir of dissolved nutrients in oceans is depleted and replenished
- the carbon cycle.

Exercises

Exercise 3.1 Marine interactions and biochemistry

1   Table 3.1 shows the productivity of different marine habitats.

Marine habitat		Annual biomass / 10^6 tonnes biomass yr^{-1}	Area / 10^6 km^2	Productivity / tonnes biomass km^{-2} yr^{-1}
coral reef		100	0.1	1 000
open ocean	tropics	11 400	190	
	temperate	22 000	100	
	polar	1 560	52	
continental shelf	no upwelling		27	400
	upwelling		0.4	980

Table 3.1: The productivity of different marine habitats.

a i   Calculate the productivity for each of the open ocean habitats.
   ii Suggest why polar waters have a lower productivity than temperate waters.
   iii Explain why coral reefs, despite having the highest productivity, are not the marine habitat that contributes the highest annual biomass.
b i   Calculate the annual biomass for each of the continental shelf habitats.
   ii Suggest why continental shelfs with upwellings have a higher productivity than those with no upwellings.
c Nursery habitats are where many juvenile species feed and grow until they are large enough to live in the open ocean. Table 3.2 shows the percentage contribution of different nursery habitats to the productivity of adult marine organisms.

Marine habitat	% contribution of juvenile habitats to adult productivity
estuaries	27
seagrass beds	31
muddy shore	11
sandy shore	3
rocky shore	4
coral reef	24

Table 3.2: Contribution of different nursery habitats to adult productivity.

   i   Calculate the average percentage contribution.
   ii Discuss why nursery habitats with the higher primary productivity contribute more to adult productivity.

2 a Copy and complete the word equation below to describe how marine producers use photosynthesis to capture the energy of sunlight to convert inorganic nutrients into organic glucose.

$$\underline{} + \underline{} \rightarrow \underline{} + \underline{}$$
$$6CO_2 + 6H_2O \rightarrow C_6H_{12}O_6 + 6O_2$$

   b A study of the diet of a seals investigated the stomach contents of nine seals. The scientists discovered a variety of invertebrate prey had been consumed: 73 crabs, 55 clams, 47 snails, 32 amphipods and 18 shrimps. Calculate the mean number of invertebrates eaten per seal. Show your working.
   c The glucose produced by photosynthesis and chemosynthesis is used in aerobic respiration to provide metabolic energy in the form of ATP. Copy and complete the following word equation.

$$\underline{} + \underline{} \rightarrow \underline{} + \underline{} + \text{cellular energy}$$
$$C_6H_{12}O_6 + 6O_2 \rightarrow 6CO_2 + 6H_2O + 32ATP$$

   d Explain how increased photosynthesis can lead to a higher **primary productivity**.

> **TIP**
> A good way to answer Question 2b would be to form a table of the similarities and differences between the energy sources, substrates and products, marine organisms and habitat.

> **KEY WORD**
> **primary productivity:** the rate of production of new biomass through photosynthesis or chemosynthesis

Exercise 3.2 Feeding relationships

This exercise will build your confidence in the use of the feeding relationships terms that underpin marine ecology, including **trophic level**, consumer, producer, herbivore, carnivore, omnivore, decomposer, predator and prey. We will then investigate food chains and food webs through the context of Antarctic and kelp forest ecosystems.

> **KEY WORDS**
>
> **trophic level:** the position an organism occupies in the food chain or food web
>
> **niche:** the role of a species within an ecosystem

1 a Explain the meaning of each of the following terms used in ecology:
   - i producer
   - ii consumer
   - iii decomposer.

  b Read the passage below and form a list of the organisms whose **niches** are producers, consumers or decomposers.

  Manatees are a type of sea cow that feed in coastal waters on a wide variety of macroalgae, as well as marine plants such as turtle rass, manatee grass, shoal grass, mangrove leaves and water hyacinths. Few predators are large enough to attack a fully grown manatee but they can be occasionally predated upon by jaguars, crocodiles and sharks. If a manatee dies in open water it sinks to the ocean floor. The first species to eat the decaying flesh of the carcass are sharks, hagfish and amphipods. Crabs, snails and worms then eat the organic leftovers in the sediment. When only the skeleton remains, bacteria break down the oils in the bones to enrich the surrounding sediments with inorganic nutrients.

  c Detritivores feed on dead and decaying material (detritus). Name three detritivores from the passage above.

2 a Using marine examples, explain the differences in meaning among the terms in each set:
   - i primary, secondary, tertiary and quaternary consumer
   - ii herbivore, carnivore and omnivore
   - iii predator and prey
   - iv food chain and food web.

  b The Antarctic is encircled by the Southern Ocean, which forms over 20% of the world's ocean area and includes the Earth's largest current, the Antarctic circumpolar current. The food webs of the Southern Ocean are among the most important in the world. They support a wide range of organisms from algae to large animals such as whales, seals and penguins, as shown in Figure 3.1.

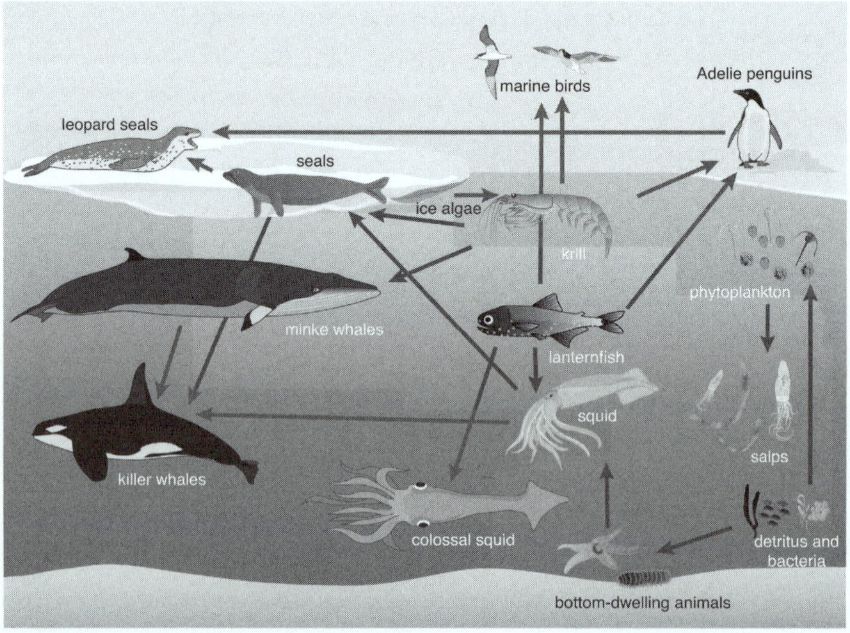

Figure 3.1: Antarctic food web.

   - i Draw a food chain that includes krill and has a quaternary consumer.
   - ii Explain what effect a decrease in krill abundance would have on the Antarctic food web.

iii Killer whales are predators of squid. Explain how the populations of the two species may be interrelated.

3 Figure 3.2 shows a kelp forest food web.

a What is the trophic level occupied by the kelp?

b Kelp and sea otters are both key species in this kelp forest ecosystem. Using Figure 3.2, copy and complete these food chains to describe examples of different trophic relationships that can link these two species.

i kelp → herbivorous fish → kelp crab → _____ → sea otter

ii kelp → invertebrate → _____ → sea otter

iii kelp → _____ → sea otter

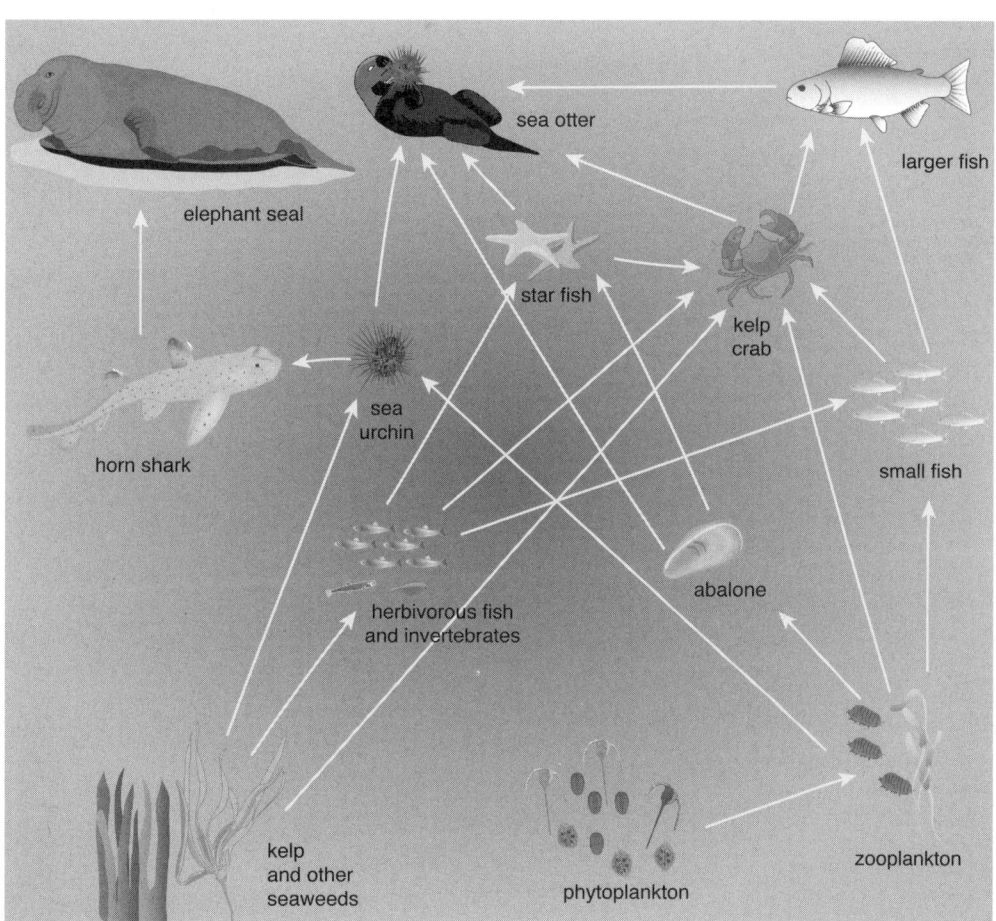

Figure 3.2: A kelp forest food web.

> **TIP**
>
> Trophic levels are given as numbers. For example, a primary consumer is described as being in the second trophic level.

c A quinary consumer is a fifth-order consumer.

i At which trophic level is a quinary consumer?

ii State an example of a food chain in Figure 3.2 where sea otters are a quinary consumer.

4 Figure 3.3 shows the changes in kelp density (graph a), grazing intensity by sea urchins (graph b), sea urchin density (graph c) and sea otter abundance (graph d) over a 13-year period from 2000 to 2012.

  a Explain why the grazing intensity was greater in 2012 than in 2002.

  b Suggest why the sea urchin population density is lower in 2002 despite there being more kelp to feed on in 2002.

  c Suggest an action that coastal communities could take to increase the local sea otter population.

  d Suggest what effect a rise in sea otter numbers could have on the horn shark population.

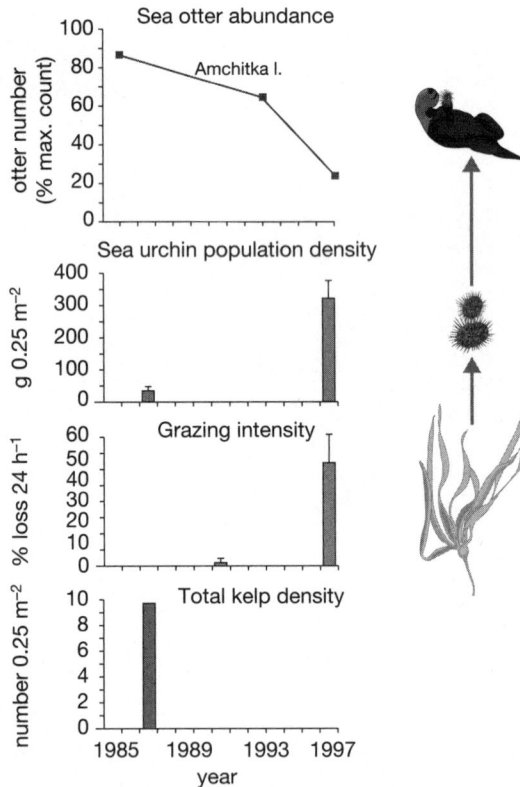

Figure 3.3: Population density of organisms in a kelp forest ecosystem.

5 Figure 3.4 contrasts the populations of kelp and sea urchins at two locations: Amchitka and Shemya Islands in the Bering Sea near Alaska.

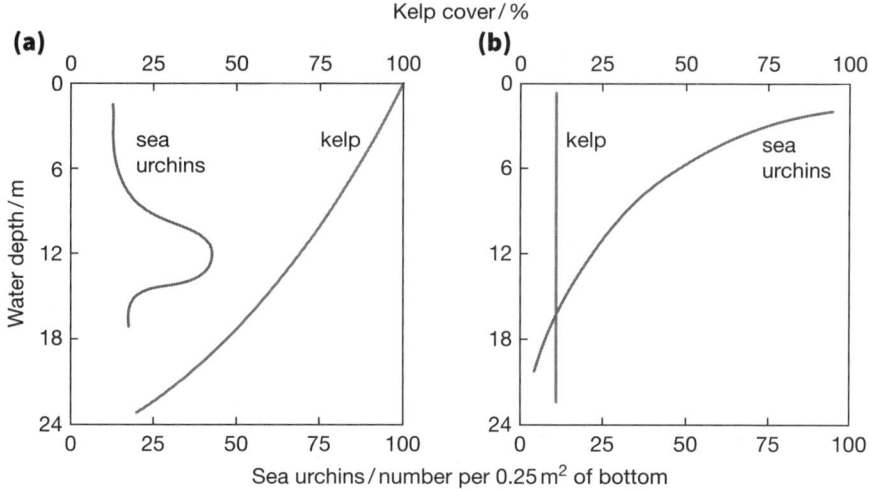

Figure 3.4: Kelp and sea urchins at (a) Amchitka Island and (b) Shemya Island.

a On Amchitka Island, why is there less kelp with increasing depth?
b At a water depth of 6 metres, state the abundance of sea urchins on Amchitka and Shemya Islands.
c At a water depth of 6 metres, state the abundance of kelp on Amchitka and Shemya Islands.
d Sea otters are abundant on Amchitka Island but rare on Shemya Island. Explain how the relative abundance of sea otters could explain the differences in kelp and sea urchin populations between the two islands.
e Predict the difference in sea otter diet between Amchitka and Shemya Islands.

Exercise 3.3 Feeding calculations and pyramids

This exercise focuses on how to calculate primary productivity (GPP and NPP) and the energy losses along food chains (TLTE). We will also study how to draw, describe and interpret the three types of food pyramids: energy, numbers and biomass.

1 **Trophic level transfer efficiency (TLTE)** measures the amount of energy that is transferred between trophic levels. TLTE is calculated by:

$$\text{trophic level transfer efficiency (TLTE)} = \frac{\text{energy trophic level}^{n+1}}{\text{energy trophic level}^{n}} \times 100\%$$

in which trophic level $n+1$ is the trophic level above trophic level n.

a Calculate the trophic level transfer efficiency (TLTE) if the secondary consumer has 500 Kcal and the primary consumer has 4500 Kcal.
b Figure 3.5 shows a pyramid of energy for a marine ecosystem (in arbitrary units). Calculate the TLTE for:
   i   phytoplankton → zooplankton
   ii zooplankton → herring
   iii mackerel → tuna

KEY WORD

trophic level transfer efficiency (TLTE): measures the amount of energy that is transferred between trophic levels

TIP

Make sure to use the correct units to refer to the rate of energy produced per area: $kJ/m^2/year$ or $kJ\,m^{-2}\,year^{-1}$.

c Using marine examples, describe why the TLTE varies among steps in a food chain.

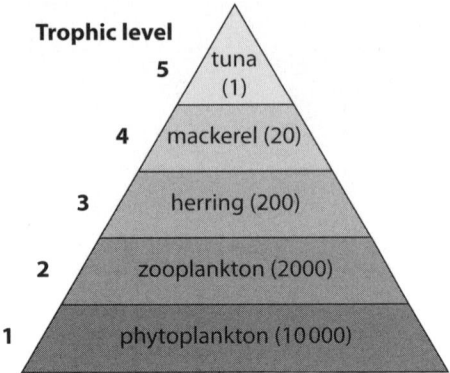

Figure 3.5: A pyramid of energy for a marine ecosystem.

2 Figure 3.6 shows an energy flow diagram for a marine ecosystem.

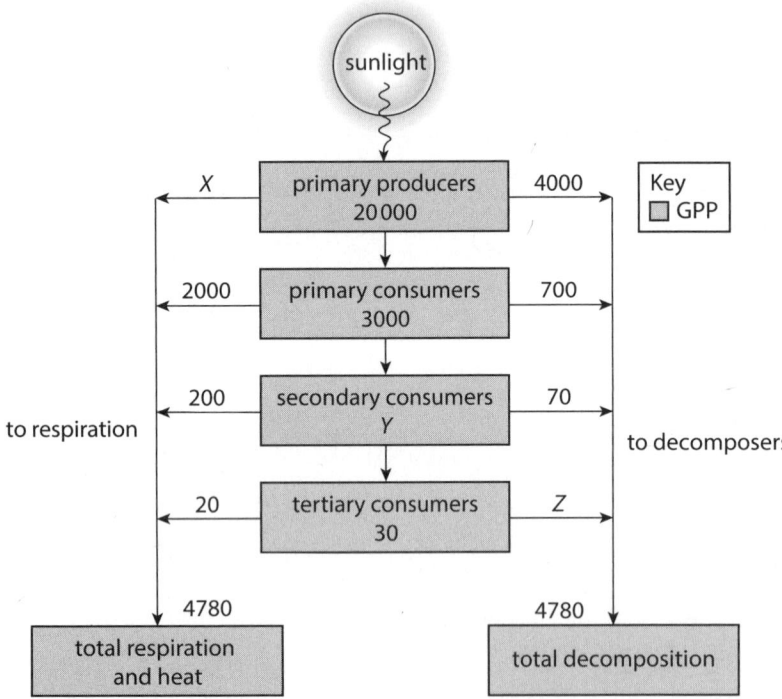

Figure 3.6: An energy flow diagram for a marine ecosystem.

a Calculate the missing values X, Y and Z.
b Calculate the total energy used in respiration.
c Calculate the trophic level transfer efficiency (TLTE) between the primary and secondary consumers.
d Calculate the energy efficiency for photosynthesis.

3   The food chain below includes data collected from a field trip to a mangrove estuary.

	Mangrove →	Prawn →	Fish →	Heron
Numbers	1	31	7	1
Biomass (g m^{-2})	800	40	10	1

a   Draw a **pyramid of numbers** for the mangrove food chain.

> **KEY WORDS**
>
> **pyramid of numbers:** a diagram that shows the number of organisms in each trophic level of a food chain
>
> **pyramid of biomass:** a diagram that shows the biomass present in each trophic level of a food chain
>
> **pyramid of energy:** a diagram that shows the amount of energy in each trophic level of a food chain

> **TIP**
>
> Remember that the overall area of the plot should aim to take up at least 50% of the graph area. For example if there were 80 lines on the y-axis of the graph paper, you could divide this into four trophic level rows of 20 lines each.

Step 1: Shape – The trophic level rows should be drawn as rectangles, not triangles. When you plot a pyramid, the bottom row is always the producer while the top row is the end of the food chain.

Step 2: y-axis – Choose the correct scale for the height of each row on the pyramid of numbers. There are four organisms in the mangrove food chain, so you will need to have four equally sized trophic level row heights that take up as much of the total y-axis as possible.

Step 3: x-axis – Choose a scale for the x-axis to fill as much of the width of the graph as possible.

Step 4: Each trophic level row must be correctly labelled with species name.

b   Draw a **pyramid of biomass** for the mangrove food chain.

c   Sketch a **pyramid of energy** for the mangrove food chain.

> **TIP**
>
> A sketch is a rough graph as there is no data provided to plot for this pyramid. You need to repeat steps 1–4 while assuming that the trophic efficiency is approximately 10%.

4   Sketch pyramids of numbers, biomass and energy for the food chain data from a kelp ecosystem in Table 3.3.

Trophic level	Label	Numbers	Biomass	Energy
tertiary consumer	chinook salmon	1	5	50
secondary consumer	slimy sculpin	10	10	400
primary consumer	molluscs	50	40	4000
producer	kelp	200	800	20000

Table 3.3: Food chain data from a kelp ecosystem.

> **TIP**
>
> Repeat steps 1–4 as above but with data for biomass.

Exercise 3.4 Nutrient cycles

This exercise will start by introducing the different types of marine nutrients (gases, inorganic ions and organic compounds) and the elements that are essential for building small organic molecules into larger ones such as carbohydrates, lipids and proteins. We will then investigate the processes by which the reservoir of dissolved nutrients in oceans is depleted and replenished as well as how a lack of nutrients can be a limiting factor to productivity. These biological, chemical, physical and geological processes will then be explored in the context of the carbon cycle.

1   a   State biochemical function for each of the following elements:
        i   nitrogen
        ii   carbon
        iii calcium
        iv   phosphorous
        v   magnesium.
    b   Name two other elements found in all carbohydrates, lipids and proteins.
    c   Which additional element is needed in proteins to form some amino acids, such as cysteine and methionine?
    d   The elements listed above are needed in relatively large amounts for productivity. Name an inorganic ion that is only needed in far smaller concentrations.
    e   Explain why the concentrations of nitrate and phosphate ions are crucial for a healthy marine ecosystem.

2   a   Nutrient cycles involve the movement and recycling of chemical elements between the **abiotic** phase (when they are inorganic chemicals) and the **biotic** phase (when they are organic chemicals). Name the process that converts chemicals in each change between phases:
        i   abiotic to biotic phase
        ii   biotic to abiotic phase.
    b   Draw a table to put the following inorganic and organic nutrients into these categories: inorganic ion, inorganic gas and organic compound.

    Mg^{2+} CO_3^{2-} lipid HCO_3^- NO_3^- Ca^{2+} NH_4^+ PO_4^{3-} NO_2^- SO_4^{2-} N_2
    SO_2 protein NO_2 CO_2 carbohydrate CO NO NH_3 S^{2-}

3   a   In the biotic phase, organic chemicals form small molecules called **monomers** that are the building blocks that can then be assembled into larger molecules called **polymers**. Name the monomers that are the sub-units of polymers of proteins, carbohydrates and lipids.
    b   Name the biochemical process that converts:
        i   monomers into polymers
        ii   polymers into monomers.
    c   Name the covalent bonds that are formed when these polymers are created:
        i   protein
        ii   lipid
        iii complex carbohydrate.

> **KEY WORDS**
>
> **abiotic:** non-living components of the ecosystem where chemicals are inorganic
>
> **biotic:** living components of the ecosystem where chemicals are organic
>
> **monomer:** the smallest unit of a polymer; monomers are able to join chemically to form longer molecules
>
> **polymer:** a large molecule made from many repeating sub-units

4   Many nutrient cycles share common processes that help cycle nutrients between the biotic and abiotic phases. Figure 3.7 shows a diagram of the carbon cycle.

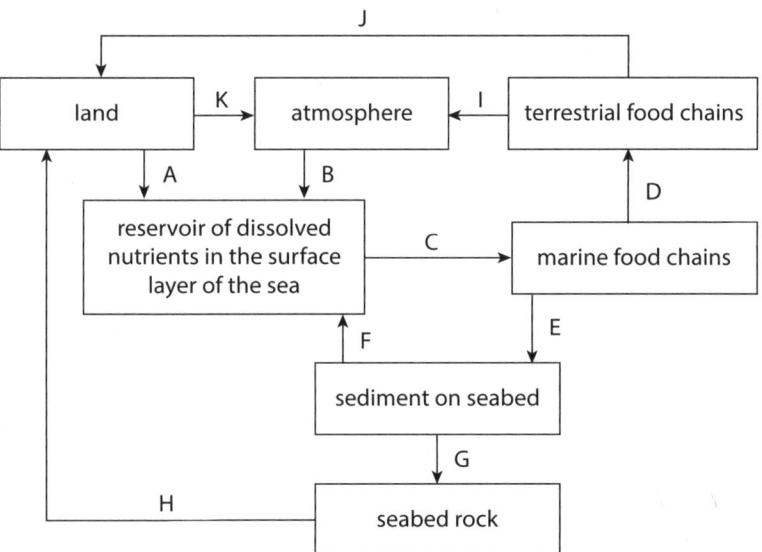

Figure 3.7: The carbon cycle.

Select from the list of biological, chemical, physical and geological processes in Table 3.4 to describe A–K on Figure 3.7.

Biological processes	Chemical processes	Physical processes	Geological processes
uptake/ assimilation	dissolving of atmospheric gases	weathering of rock	volcanic activity
photosynthesis		erosion of rock	formation of rock
respiration	combustion	run off	plate tectonics (uplifting)
evapotranspiration		upwelling	
harvesting (Fishing)		precipitation	
decomposition			
detritus / marine snow			

Table 3.4: Common processes in nutrient cycles.

5. Inorganic ions dissolved in the surface water of the ocean form a reservoir of soluble nutrients available to producers and consumers.

   a Explain how each of the following processes can replenish the inorganic nutrients used to create new biomass:

       i   upwelling
       ii run off
       iii tectonic activity
       iv dissolving of atmospheric gases.

   b Explain how and why nutrients can be depleted from the ocean reservoir by:

       i   uptake into organisms
       ii harvesting
       iii formation of rocks.

EXAM-STYLE QUESTIONS

1 Figure 3.8 shows a marine food web.

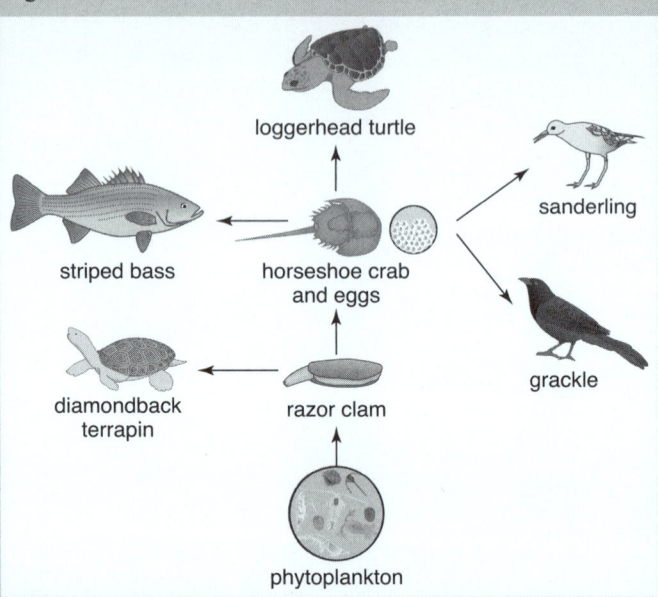

Figure 3.8: A marine food web

a	What is the primary source of energy for this food web?	[1]
b	Explain what is meant by *trophic level*.	[1]
c	Write a food chain including a tertiary consumer.	[2]
d	Explain what the arrows between organisms represent.	[2]
e	With reference to Figure 3.8, explain the term *predator*.	[2]
f	Suggest *one* biotic factor, other than predation, that might affect the horseshoe crab population.	[1]
g	Ecologists studied a total of 130 horseshoe crabs and found 10 with ciliate parasites. Explain the term *parasitism*.	[3]

[Total: 12]

CONTINUED

2 Figure 3.9 shows a food web for the Arctic.

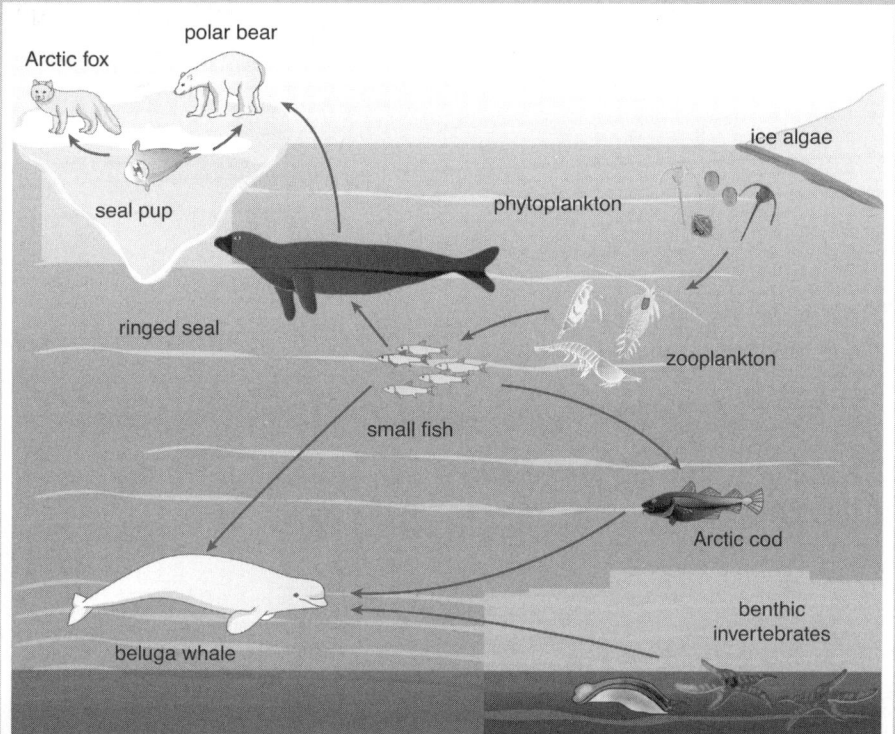

Figure 3.9: An Arctic food web

   a   Using Figure 3.9, state a food chain with polar bears as a quaternary consumer. [1]

   b   A study of the diet of a seals investigated the stomach contents of nine seals. The scientists discovered a variety of invertebrate prey had been consumed: 73 crabs, 55 clams, 47 snails, 32 amphipods and 18 shrimps. Calculate the mean number of invertebrates eaten per seal. Show your working. [2]

   c   Beluga whales are hunted in Arctic regions for their meat, blubber and skin. Explain how this could directly affect the population numbers of each of the other species in the following food chain: [3]

   phytoplankton → zooplankton → small fish → beluga whale

   d   Draw a pyramid of energy for the food chain above. [4]

   [Total: 10]

3 a   Define the term *nutrient*. [2]

   b   Explain the difference between the terms *biomass* and *productivity*. [2]

   c   A major limiting factor for phytoplankton growth in Antarctic waters is dissolved iron. Why might the concentration of iron be relatively high in spring? [2]

CONTINUED

d Figure 3.10 compares biological pyramids between two marine food chains. **Justify** why the pyramids for the two food chains are either different or the same. [6]

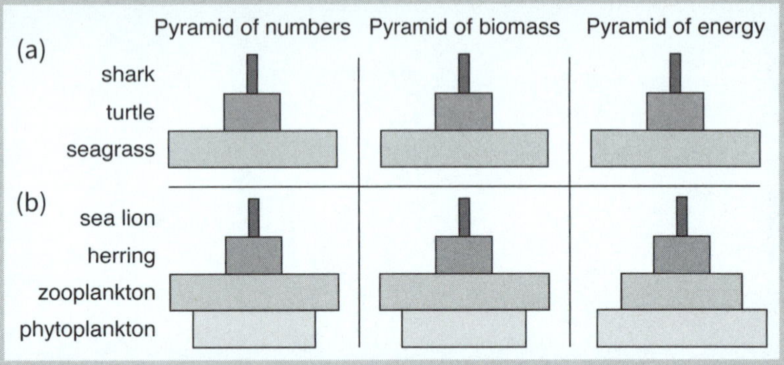

> **COMMAND WORD**
>
> **justify:** support a case with evidence / argument

Figure 3.10: Antarctic food pyramids

[Total: 12]

4 Figure 3.11 shows a diagram of the carbon cycle.

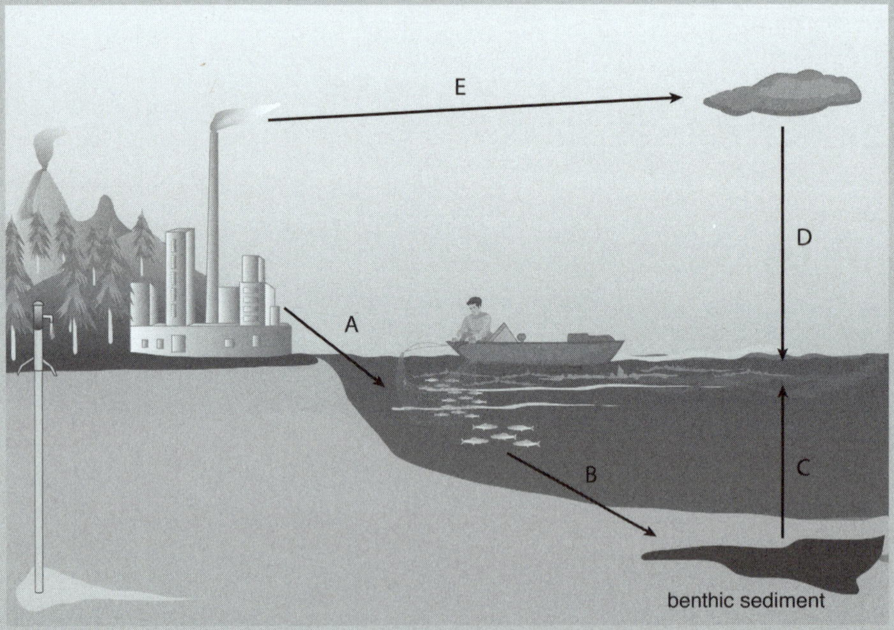

Figure 3.11: The Carbon cycle

a Name the processes labelled A–E. [5]

b Explain how the following abiotic factors can affect the exchange of carbon dioxide gas between the atmosphere and surface waters:

    i increased wind speed [2]

    ii increased ocean temperature. [2]

c Explain *two* biological processes by which marine microbes recycle carbon between the organic nutrient in the biotic phase and the inorganic nutrient in the abiotic phase. [4]

[Total: 13]

Chapter 4
Classification and biodiversity

CHAPTER OUTLINE

The questions in this chapter cover the following topics:
- the use of binomial nomenclature, taxonomic hierarchies and dichotomous keys
- the drawing and labelling of biological specimens
- the adaptive features and ecological importance of marine producers
- the economic and ecological importance of cartilaginous fish
- the importance of maintaining marine biodiversity
- how to perform safe and ethical fieldwork
- building your confidence with maths.

Exercises

Exercise 4.1 Classification

This exercise will start by using some of the keystone species detailed in the syllabus to build your confidence in the use of binomial nomenclature as part of the taxonomic hierarchies. We will then practise the use and design of dichotomous keys and biological drawings.

1 Each **species** has a unique name comprised of its genus and species. These two names are called **binomial nomenclature** and are part of the **taxonomic hierarchy**.

   a Copy and complete Table 4.1 for each of the species.

Taxonomic hierarchy	Turtle grass	Giant kelp
		Eukarya
Kingdom		
	Angiospermaphyta	Phaeophyta
Class	Magnoliopsida	Phaeophycae
	Alismatales	Laminarialis
Family	Hydrocharitaceae	Lessoniaceae
	Thalassia	
	testudinium	

Table 4.1: Taxonomic hierarchy for turtle grass and giant kelp.

KEY WORDS

species: a group of similar organisms that can interbreed naturally to produce fertile offspring

binomial nomenclature: the two-part Latin name given to each species comprising the genus followed by the species

taxonomic hierarchy: the classification of the species within living organisms by describing the domain, kingdom, phylum, class, order, family, genus and species

4 Classification and biodiversity

> **TIP**
>
> When you are writing the biological name for a species, the genus starts with a capital letter while the species is in lower case. When you write them names by hand, it is best practice to underline both genus and species.

> **KEY WORD**
>
> **dichotomous key:** an identification tool utilising a series of choices between alternative characters, with a direction to another stage in the key, until the species is identified

b Use the **dichotomous key** in Table 4.2 and the taxonomic hierarchy in Table 4.3 to find the common names of species A, B and C.

Step	Characteristic	Common name
1a	Animal … Go to 2.	
1b	Not an animal …. Go to 3.	
2a	Chordata (vertebrate … Go to 4.	
2b	Non chordate (invertebrate) … Go to 5.	
3a	Plantae	turtle grass
3b	Protoctist	giant kelp
4a	Mammal … Go to 6.	
4b	Fish … Go to 7.	
5a	Echinoderm … Go to 8.	
5b	Arthropod	Antarctic krill
6a	Genus *Orcinus*	killer whale
6b	Genus *Delphinus*	common dolphin
7a	Class Chondrichthyes	blue shark
7b	Class Actinopterygii	Peruvian anchoveta
8a	Order Acanthasteridae	crown of thorns starfish
8b	Order Valvatida	bat star

Table 4.2: Dichotomous key.

Taxonomic hierarchy	Species A	Species B	Species C
Domain	Eukarya	Eukarya	Eukarya
Kingdom	Animalia	Animalia	Animalia
Phylum	Chordata	Chordata	Echinodermata
Class	Mammalia	Mammalia	Asteroidea
Order	Cetecea	Cetecea	Valvatida
Family	Delphinidae	Delphinidae	Oreasteridae
Genus	*Orcinus*	*Delphinus*	*Asterina*
Species	*orca*	*delphis*	*miniata*

Table 4.3: Taxonomic hierarchy for species A, B and C.

c   Copy Table 4.4. Use the **biological drawings** of the six species of shark in Figure 4.1 to complete the dichotomous key.

> **KEY WORD**
>
> **biological drawing:** a scientific drawing that records an image and important features of the specimen

hammerhead shark
Sphyrna zygaena

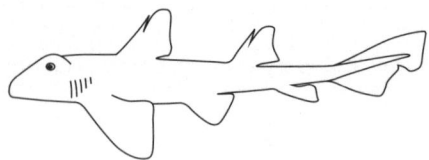
horn shark
Heterodontus francisci

goblin shark
Mitsukurina owstoni

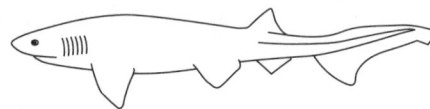
bluntnose sixgill shark
Hexanchus griseus

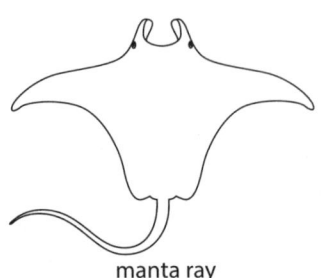
manta ray
Cephalopterus manta

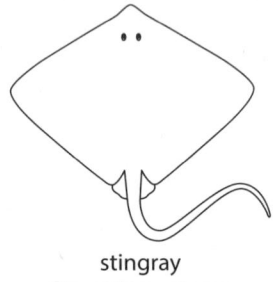
stingray
Dasyatis pastinaca

Figure 4.1: Six different shark species.

Step	Characteristic	Binomial name	Common name
1a 1b	kite-like shape to body … Go to 2. non kite-like shape to body … Go to 3.		
2a 2b	horn-like appendages on snout no horn-like appendages on snout		
3a 3b	6 gill slits 5 gill slits … Go to 4.		
4a 4b	spines on dorsal fin no spines on dorsal fin … Go to 5.		
5a 5b	long point on end of snout no long point on end of snout ….Go to 6.		
6a 6b	big eyes surrounded by a small ring eyes on end of hammer-like projection		

Table 4.4: Dichotomous key for species of shark.

4 Classification and biodiversity

2 a Draw a biological drawing of the crustacean shown in Figure 4.2.
Add the following labels: carapace, segmented abdomen, jointed legs, two pairs of antennae.

Figure 4.2: Australian red claw crayfish.

TIP

It is important to get the correct scale in a biological drawing. For example the head of the crayfish in Fig 4.2 should be drawn approximately one quarter of size of the total body length.

Biological drawings should not be coloured. Use a sharp HB pencil to draw bold lines without shading or gaps.

Step 1: Draw the outline of the crayfish, making sure that it is greater than 50% of the size of the space provided.

Step 2: Check that the diagram is biologically correct. You should have the correct numbers of claws (2), antennae (4 small, 2 large), walking legs (8) and abdominal segments (5).

Step 3: Add labels on either side of the diagram. You can write them in ink, but you should draw the label lines in pencil using a ruler.

b Make a copy of Figure 4.3 and add annotated labels to name and describe the function of the parts of the peacock mantis shrimp: tail fan, walking legs, swimming legs, claw, carapace, antennae.

Figure 4.3: Peacock mantis shrimp.

TIP

Remember that annotated labels do not simply name the part but also give a function for it: *tail fan: assists the crayfish in moving backwards.*

3 Figure 4.4 shows a diagram of a Placoderm, a prehistoric fish that could grow to over 6 metres in length.

Figure 4.4: Placoderm.

a   Make a copy of Figure 4.4 and add the following labels:
   - pectoral fin
   - pelvic fin
   - heterocercal tail fin
   - dorsal fin
   - bony plates on head and thorax.

b   Suggest an advantage of the feature that is not found in most modern fish species.

Exercise 4.2 Marine organisms

This exercise will help build your understanding of two of the key groups of marine organisms that are covered in Chapter 4. We will begin with marine producers, examining their key similarities and differences, their adaptive features and how they promote biodiversity. We will then explore the ecological and economic importance of cartilaginous fish.

1   a   Create a table to compare three groups of marine producers: phytoplankton, macroalgae and marine plants. You should include details of the kingdoms they are classified under as well as the names of common examples.

    b   Create a table to compare the differences between macroalgae and marine plants that can be observed in Figure 4.5.

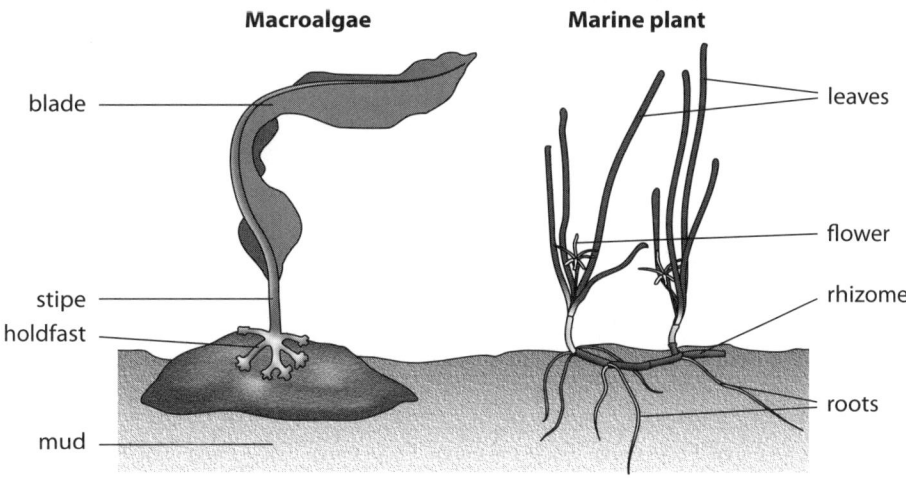

Figure 4.5: Macroalgae and marine plants.

2 a Seagrasses are marine plants that inhabit shallow clear coastal waters. Discuss how each of the following **abiotic factors** is most likely to affect the growth of seagrass:
  i temperature
  ii **turbidity**
  iii depth of water.

 b Seagrass habitats vary from small clumps of a single seagrass species to extensive multispecies meadows covering large areas of the seabed. Describe how the leaves and roots of seagrasses can help maintain water quality with regard to:
  i turbidity
  ii **marine toxins**.

 c State one way in which each of the following activities may decrease biodiversity by damaging the delicate ecological balance of marine plants in a coastal ecosystem.
  i run-off from cities
  ii nutrients leached from farmland
  iii power stations
  iv boat owners
  v fishermen.

 d **Biotic factors** can also destroy the fragile habitat of a seagrass meadow. Predict the result of each of the following events:
  i In the early 1930s a wasting disease caused by the *Labyrinthula* slime mould devastated 90% of all eelgrass (*Zostera marina*) growing in temperate North America. What effect would this have on the populations of *Lottia alveus*, a limpet specialised to live on eelgrass?
  ii In the 1980s *Caulerpa taxifolia*, an **invasive species** of ornamental seaweed found in aquaria, was released into Mediterranean seagrass meadows dominated by the native Neptune seagrass (*Posidonia oceanica*). Suggest why the invasive species proved detrimental to the Mediterranean's seagrass ecosystem.

KEY WORDS

abiotic factors: the environment's geological, physical and chemical features; the non-living part of an ecosystem

turbidity: the level of transparency loss water has due to the presence of suspended particles in the water; the higher the turbidity, the harder it is to see through the water

marine toxins: poisonous chemicals that can contaminate seawater

biotic factors: the living parts of an ecosystem, which includes the organisms and their effects on each other

invasive species: species that have become established in an area that is not their normal habitat due to human activity

3   The carbon stored in marine ecosystems is known as '**blue carbon**' because it is stored in the sea.

    a   State the cellular processes by which marine producers can both generate and store carbon dioxide.

    b   It has been estimated that in this way the world's seagrass meadows can capture up to 83 million metric tonnes of carbon each year. While seagrasses occupy only 0.1% of the total ocean floor, they are estimated to be responsible for up to 11% of the organic carbon buried in the ocean. Calculate the total annual amount of carbon captured by the World Ocean.

    c   One square metre of seagrass can absorb 83 g of carbon per year. If an average car emits 6500 kg of carbon per year, what area of seagrass would be needed to act as a carbon sink (a site that absorbs more net carbon dioxide than it releases) for it?

> **KEY WORD**
>
> **blue carbon:** carbon stored in marine ecosystems

4   a   Table 4.5 shows the global catch of cartilaginous fish between 1985 and 2000.

Year	1986	1988	1990	1992	1994	1996	1998	2000
Catch (thousands of tonnes)	634	693	693	729	757	814	816	828

Table 4.5: Global catch of cartilaginous fish between 1986 and 2000.

> **TIP**
>
> Remember to include your working and units in your answer.

> **TIP**
>
> This question asks for a histogram rather than a bar chart. Check that you know the difference.

    i   Plot a histogram of cartilaginous fish caught between 1986 and 2000.

    ii Calculate the percentage increase in catch between 1986 and 2000. Give your answer to the nearest whole number.

    b   Figure 4.6 shows the population of gray reef and silver tip sharks in the Chagos archipelago in the Indian Ocean between 1976 and 2012.

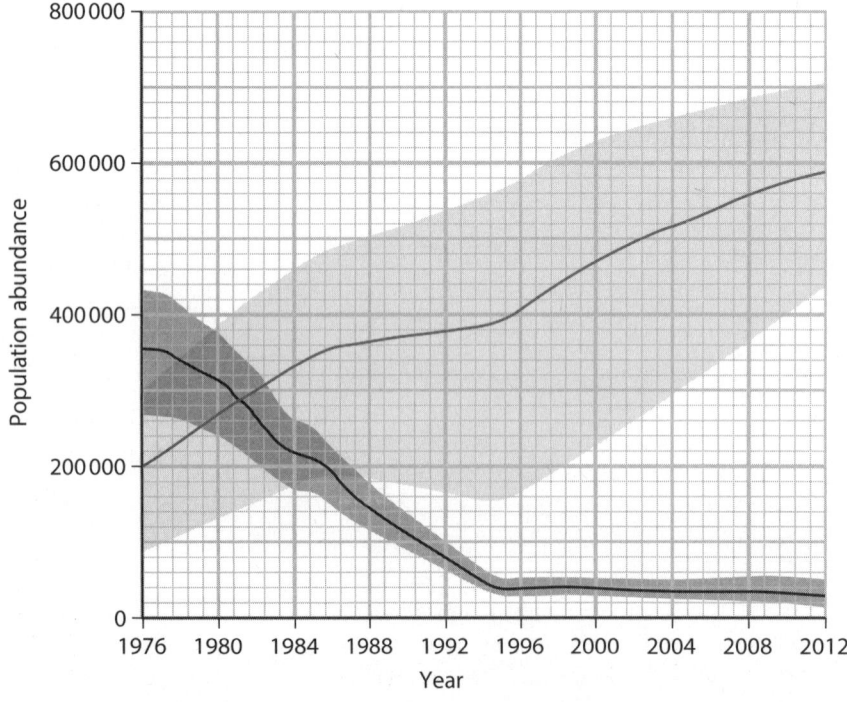

Figure 4.6: Population of gray reef and silver tip sharks from 1976 to 2012.

> i   Describe the change in population of the silver tip shark population between 1976 and 2012.
> ii Suggest the most likely reasons for the change in silver tip shark population.
> iii Suggest a reason for the change in gray reef shark population.
> iv Predict how the change in shark populations may have impacted the marine biodiversity of the Chagos archipelago.

> **TIP**
>
> Support your answer with manipulated data. That is, do a quick calculation of the change rather than simply stating values directly from the table.

Exercise 4.3 Biodiversity

This exercise will explore the different levels of biodiversity and why maintaining biodiversity is important. We will then look at different methods of sampling populations and how to perform safe and ethical fieldwork.

1 a Define biodiversity.
   b Using rocky shores as an example, explain the three different levels that can define biodiversity.
   c Using coral reefs as an example, explain the importance of marine biodiversity with regard to:
      i   maintaining stable ecosystems
      ii protecting the physical environment
      iii climate control.

2 Lake Hodgson in Antarctica has been sealed beneath 4 m of ice for at least 11 000 years. Analysis of sediment taken from beneath the subglacial waters has shown that about one-fourth of the DNA present comes from previously unknown species.
   a Suggest different methods by which research scientists could sample organisms living on the **benthic floor**.
   b Explain the potential therapeutic benefit from these unknown marine organisms.

3 a An **ethical issue** is when a person has to choose between alternatives that can be evaluated as either right (ethical) or wrong (unethical). Suggest a benefit of each of these following ethical approaches to sampling on a rocky or sandy shore.
      i   treading carefully
      ii   handling creatures carefully with wet hands
      iii putting creatures back in the same location where you found them
      iv   limiting your collection of empty shells
      v   picking up any litter found
      vi   leaving attached seaweed in place
      vii working in pairs and pooling class data
      viii staying away from breeding birds.

> **TIP**
>
> The context of this question is the benthic floor, so you must give an answer that relates to this specific habitat.

> **KEY WORDS**
>
> **benthic floor:** the habitat at the bottom of the ocean
>
> **ethical issue:** when an individual or group has to choose between alternatives that can be evaluated as either right or wrong

b Safety is also paramount on any field investigation. For each of these following potential hazards, describe why it is dangerous and a way to reduce its risk:
   i   heavy rain
   ii cold weather
   iii hot weather
   iv tide times
   v   footwear
   vi breeding sea lions
   vii working alone
   viii face the ocean
   ix rock surface.

c List any other safety measures that should be taken by schools when planning a marine science field trip.

Exercise 4.4 Maths skills: biomaths

This exercise will help to build your confidence in using the range of mathematical tools that are introduced in this chapter. We will cover the **Lincoln index**, **Simpson's index of diversity (D)** and **Spearman's rank correlation (r_s)**.

1 The Lincoln index is one of a number of mathematical equations that can be used to estimate population size of a **motile species** using the following formula and symbols:

$$N = \frac{n_1 \times n_2}{m_2}$$

N = total population size
n_1 = number of specimens captured and marked on the first visit
n_2 = number of specimens captured (both marked and unmarked) on the second visit
m_2 = number of specimens recaptured that were marked.

a Black nerites, *Nerita atramentosa*, are medium-sized sea snails that are commonly found on intertidal rocky shores. A class estimated the population on a shore close to their school using the **mark–release–recapture** method. Estimate the population size (N) of black nerites using the following fieldwork data.

mark–release–recapture method		Fieldwork data	Total population size (N)
n_1	number of black nerites captured and marked on the first visit	25	
n_2	number of black nerites captured (both marked on unmarked) on the second visit	40	
m_2	number of recaptured black nerites that were marked	10	

b A biologist wants to estimate the size of a population of hammerhead sharks in the coastal waters of an island. She captures 20 hammerhead sharks on her first visit, and she marks their dorsal fin with an easily visible numbered tag. One month later she returns to the island and looks for tags on each hammerhead shark that she observes. Five of the ten sharks observed have tags on their dorsal fins.

KEY WORDS

Lincoln index: a mathematical equation that can use the mark–release–recapture data to estimate the population size

Simpson's index of diversity (D): a biodiversity measure that accounts for both species richness and evenness

Spearman's rank correlation (r_s): a mathematical tool used to find out if there is a correlation between two sets of variables, when they are not normally distributed

motile species: an organism that can move around in its habitat and is not fixed in place

mark–release–recapture: a method to estimate the population size of mobile species

TIP

You cannot have a 'part of an organism', so when predicting populations the answer must be given as a whole number, you cannot have decimal places.

4 Classification and biodiversity

    i Use this data to estimate the total hammerhead shark population.
    ii Suggest reasons that the biologist marked the dorsal fins of the hammerhead sharks with easily visible numbered tags.
    iii Suggest a source of experimental error.

c The total population of salmon in a fish farm is known to be 2487. A student used the mark–release–recapture method to see how close his estimate would be to the actual stock size. The student caught 50 salmon in his first and second visits and estimated that the stock size was 1250.
    i State how many of the first and second samples were marked.
    ii Discuss why there is variance between the actual and estimated stock sizes and how this could be improved.

> **TIP**
>
> To find m_2, you will need to rearrange the Lincoln index equation:
>
> $$m_2 = \frac{n_1 \times n_2}{N}$$

2 Simpson's index of diversity (D) can be used calculate to the biodiversity found at different locations using the equation:

$$D = 1 - \left(\Sigma\left(\frac{n}{N}\right)^2\right)$$

Σ = sum of (total)
n = number of individuals of each *different* species
N = the total number of individuals of *all* the species.

a Calculate the biodiversity found at a deep-sea hydrothermal vent using Simpson's index of diversity (D).

Step 1: Copy and complete Table 4.6 to calculate N; $\frac{n}{N}$; $\left(\frac{n}{N}\right)^2$; and $\Sigma\left(\frac{n}{N}\right)^2$.

Organism	Number (n)	n/N	$\left(\frac{n}{N}\right)^2$
tube worms	400		
clams	10		
shrimps	30		
crabs	4		
Total	$N =$		$\Sigma\left(\frac{n}{N^2}\right) =$

Table 4.6: Calculating N and $\Sigma\left(\frac{n}{N^2}\right)$.

> **TIP**
>
> Simpson's index of diversity (D) can be stated to either 2 or 3 decimal places.

Step 2: Calculate Simpson's index of diversity (D).

b Table 4.7 shows population data collected as part of a marine conservation project in New Zealand. Use Simpson's index of diversity (D) to determine whether the biodiversity is greater inside or outside the **marine protected area (MPA)**.

Organism	Outside MPA	Inside MPA
spiny lobster	1	25
goatfish	12	13
snapper	4	12
spotty	10	14
red moki	7	21
blue cod	6	10
leather jacket	11	19

Table 4.7: New Zealand marine conservation project data.

> **KEY WORD**
>
> **marine protected area (MPA):** an area of ocean or coastline where restrictions have been placed on activities; the levels of restriction may vary, some may be no-take areas where no fishing is permitted, others may allow some fishing, some may ban all access to unauthorised people, while others may allow restricted access

3   The Spearman's rank correlation (r_s) is commonly used to find out whether there is a correlation between two sets of variables, when they are not normally distributed. This can be particularly useful to test the strength of the relationships between the distribution and abundance of two species or when you wish to investigate whether the abundance of a species is linked to an abiotic factor. The Spearman's rank correlation (r_s) can be calculated using the following formula and symbols:

$$r_s = 1 - \left(\frac{6 \times \Sigma D^2}{n^3 - n}\right)$$

r_s = Spearman's rank correlation
Σ = 'sum of'
D = the difference between each pair of ranked measurements
n = the number of pairs of items in the sample.

a   Calculate a Spearman's rank correlation (r_s) to investigate whether there is a correlation between the abundance of oyster borers and mussels on a rocky shore (Table 4.8).

Quadrat	1	2	3	4	5	6	7	8
oyster borer number	5	8	3	16	13	10	6	7
mussels number	22	40	12	60	28	38	25	32

Table 4.8: Abundance of oyster borers and mussels on a rocky shore.

Step 1: Sketch a scatter graph to test whether a correlation is suggested.

Step 2: Test the strength of this correlation by calculating the r_s. Start by counting the number of pairs of items (n) in the data set. Then calculate n^3.

Step 3: Rank the number of oyster borers from most to least frequent and record the order 1 to 8. The quadrat with the most oyster borers is ranked 1 and the quadrat with least oyster borers is ranked 8. Repeat this step to rank the mussel abundance.

Step 4: Calculate the differences in rank in each quadrat (D) by subtracting the rank of oyster borers from the rank of mussels. Then proceed to square each difference in rank to give D^2.

Step 5: Add all the values of D^2 to find ΣD^2.

Step 6: Calculate the Spearman's rank correlation.

Step 7: Use the critical values of r_s at 0.05 probability in Table 4.9 to test whether r_s is significant at a 5% probability level.

> **TIP**
>
> The r_s critical vales table will indicate whether your r_s should be calculated to 2 or 3 decimal places.

n (number of pairs)	8	9	10	11	12
Significance level 5%	0.738	0.700	0.648	0.618	0.618
Significance level 1%	0.881	0.883	0.794	0.755	0.727

Table 4.9: Critical values of r_s at 0.05 probability.

Step 8: State a conclusion related to the statistical significance of the difference between the critical values and the r_s.

b Use the data in Table 4.10 to calculate a Spearman's rank correlation (r_s) to investigate whether there is a correlation between depth of water and the abundance of turtle grass in a seagrass meadow.

Sample site	1	2	3	4	5	6	7	8	9	10	11	12
depth pf water / m	3.0	3.5	0.7	1.4	1.0	1.2	3.3	2.0	1.6	2.6	2.8	1.8
turtle grass density / a.u.	1	0	62	18	29	22	2	5	12	4	6	10

Table 4.10: Depth of water and the abundance of turtle grass in a seagrass meadow.

EXAM-STYLE QUESTIONS

1 Figure 4.7 shows the underneath (ventral) surface of a ray.

Figure 4.7: Ventral surface of a ray

    a Add labels to a copy of Figure 4.7 to indicate the following structures: **[4]**

        gill slits   pectoral fin   pelvic fin   caudal fin

    b Make a large drawing of the underneath (ventral) surface of a ray in Figure 4.7. No labels are required. **[4]**

> **CONTINUED**

  c   Copy Table 4.11. Fill in the eight spaces to show how a smooth butterfly ray, *Gymnura micrura*, is classified. [4]

Taxonomic hierarchy	Smooth butterfly ray
	Eukarya
Kingdom	
Phylum	
	Chondricthyes
	Myliobatiformes
	Gymnuridae
Genus	
Species	

Table 4.11: Taxonomic hierarchy for a smooth butterfly ray, *Gymnura micrura*

[Total: 12]

2 a   State *two* external features that differentiate bony fish from cartilaginous fish. [2]

   b   Name and explain the function of the specialised buoyancy organ found in bony fish. [2]

   c   Describe the function of the common features shared by bony and cartilaginous fish as members of the phylum chordata. [4]

   d   Grey nurse sharks in south-east Australian waters have been analysed using a variation of the mark–release–recapture method, called mark–release–resighting. Sharks were fed a baited, barbless hook attached to a rope. Once hooked, the sharks were slowly pulled to the surface and into a stretcher before being restrained at their head and tail. An air-powered drill was then used to attach two identical numbered tags onto their dorsal fins. The scientists also recorded other data, including markings, size, weight and sex of the shark. The sharks were then released and local scuba divers and spear fishermen were asked to report sightings of sharks, including their locations and whether they were tagged.

   i   What could be the benefits of attaching two identical tags? [2]

   ii   Apart from estimating population size, suggest why scuba divers and fishermen were asked to report the location. [1]

   iii Why were the sharks not recaptured? [1]

   iv   Suggest another way of recording tagged marine organisms that does not require recapture. [1]

[Total: 13]

CONTINUED

3   *Sargassum* is a genus of brown macroalgae that is found floating on the surface of the oceans and can travel many miles across deep ocean. A piece of *Sargassum natans* is shown in Figure 4.8.

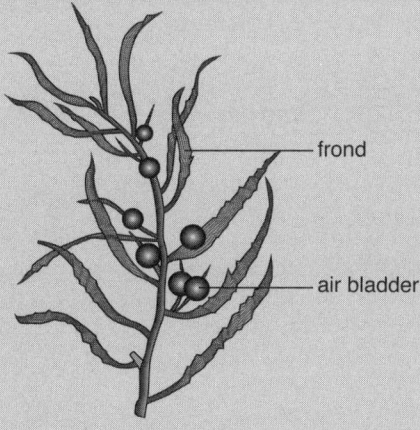

Figure 4.8: Sargassum natans

   a   Explain how *Sargassum natans* is adapted to maximise its productivity. [4]
   b   Thick mats of *Sargassum natans* are found on the surface of the Sargasso Sea where they promote biodiversity. Explain why many other species are dependent on *Sargassum natans*. [4]
   c   Fish can eat great quantities of invertebrate larvae including those of sea urchins which graze on macroalgae.
      i   Predict the ecological impact of overfishing on macroalgae abundance. [1]
      ii Macroalgae are a key habitat for lobsters. Suggest how a reduction in habitat area would affect the lobster population. [3]

[Total: 12]

4   a   Explain the importance to coastal communities of maintaining the biodiversity of coral reefs with regards to:
      i   medicine [2]
      ii   food [2]
      iii climate stability. [2]
   b   Suggest how ecologists can estimate the abundance of organisms living in a littoral zone. [6]

[Total: 12]

Chapter 5
Examples of marine ecosystems

CHAPTER OUTLINE

The questions in this chapter cover the following topics:

- the world's five oceans and calculating their mean volumes and percentage area coverage
- identifying zones found in the open ocean and the importance of oceans as carbon sinks
- conditions for tropical coral reef growth, symbiotic relationships with zooxanthellae and the difference between hard and soft coral
- the four types of tropical coral reef
- the key terms associated with coral types, structure and function
- how temperature affects reef erosion
- the use of artificial reefs
- the adaptions of organisms to living on sandy and rocky shores
- particle size and permeability on a sandy shore
- the adaptive features that allow different mangrove species to thrive in estuarine niches around the world
- the importance of, and threats to, the conservation of mangrove forests.

Exercises

Exercise 5.1 The open ocean

This exercise will increase your confidence regarding the calculation of means, percentages and percentage change.

1 The world's five oceans are interconnected and encircle the Earth as a **World Ocean**. State the name of each of the oceans A–D in Figure 5.1.

> **KEY WORD**
>
> **World Ocean:** the combination of all major oceans into one large, interconnected body of water that encircles the world's continents

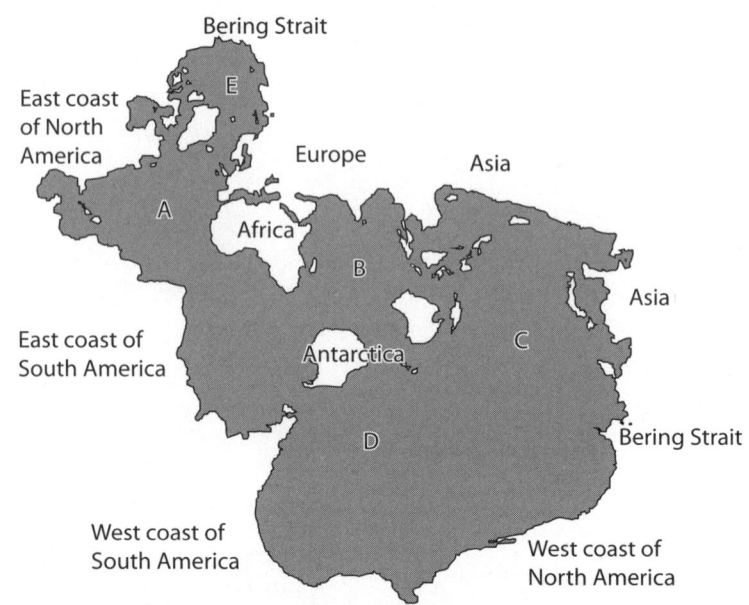

Figure 5.1 The World Ocean.

5 Examples of marine ecosystems

2 Table 5.1 shows the area and volume of each the world's five oceans:

Ocean	Area / km²	Volume / km³
Pacific	155 557 000	660 000 000
Atlantic	76 762 000	310 411 000
Indian	68 556 000	264 000 000
Southern	20 327 000	71 800 000
Arctic	14 056 000	18 750 000

Table 5.1: Area of world's oceans.

a i Use the data in the Table 5.1 to calculate the mean area of the five oceans.

Step 1: Calculate the total area covered by the world's five oceans.

Step 2: Divide the total ocean area by the number of oceans.

ii Use the data in the Table 5.1 to calculate the mean volume of the five oceans.

b i The ocean with the smallest volume is the Arctic Ocean. Using this equation, calculate how much larger in percentage is the volume of the Southern Ocean. Show your working.

$$\text{Percent increase Southern Ocean} = \frac{\text{Southern volume} - \text{Arctic volume}}{\text{Arctic volume}} \times 100$$

ii Calculate how much larger in percentage is each of the other three oceans, compared to the volume of the Arctic Ocean.

3 The area of the Pacific Ocean, as a percentage of the World Ocean, can be calculated as follows:

$$\text{Percentage area Pacific Ocean} = \frac{\text{Pacific Ocean area}}{\text{total ocean area}}$$

$$= \frac{155\,557\,000}{335\,258\,000}$$

$$= 46\%$$

a Using the data in Table 5.1, calculate the percentage area of the World Ocean covered by each of the other four oceans.

b The total area of the Earth is 510 072 000 km². Calculate the percentage area of the Earth covered by each of the oceans.

c Using the data in Table 5.1, calculate the percentage volume of the World Ocean covered by each of the five oceans.

4 Echo sounding using **SONAR** can be used to measure the depth of oceans. This can be useful when deciding where to lay submarine cables. Table 5.2 shows the mean and maximum depths for each of the world's oceans.

Ocean	Mean depth / metres	Max depth / metres
Pacific	4 080	10 994
Atlantic	3 646	8 486
Indian	3 741	7 906
Southern	3 270	7 075
Arctic	1 205	5 567

Table 5.2: Mean and maximum depths for each of the world's oceans.

> **KEY WORD**
>
> **SONAR:** a method that is used to detect underwater objects by the reflection of sound waves

a For each ocean, what percentage of its maximum depth is the mean depth?

b Calculate how much deeper in percentage is the maximum depth of the Pacific Ocean compared to the Southern Ocean. Show your working.

Exercise 5.2 The tropical coral reef

In this exercise you will gain confidence with the key terms associated with coral structure and function. You will then focus on improving your skills when tackling questions where you need to analyse data in tables or bar charts. The last question challenges you to apply your knowledge of corals to a new context.

1 a State the key word for each definition in Table 5.3.

Key word	Definition
i	a ring surrounding the mouth that helps with capturing food, expelling waste and clearing away debris
ii	the stinging cells in the epidermis of tentacles that help capture food
iii	where the polyp takes in food and expels waste
iv	where ingested food is digested
v	the protective cup within which the polyps sits
vi	the walls surrounding the calyx
vii	the base of the calyx upon which the polyp sits
viii	a single coral organism

Table 5.3: Key word definitions.

> **TIP**
>
> Key words are often used in the stem of a question to indicate what your answer should specifically relate to. You will be expected not only to understand the key words but also to be able to provide definitions for them.

b The coenosarc is a thin band of living tissue that connects individual polyps to one another. Suggest how this may be beneficial to the coral as a colonial organism.

c In the context of corals compare the following key words:
   i photosynthetic vs. heterotrophic nutrition
   ii sessile vs. motile
   iii symbiotic vs. mutualistic relationships
   iv hard vs. soft corals.

5 Examples of marine ecosystems

2 Coral cover is a measure of the proportion of reef surface covered by live stony coral rather than other organisms such as sponges, algae or starfish. Table 5.4 shows the decline in Caribbean reefs over a 20-year period.

Coral Reef	Initial coral cover / %	Coral cover 20 years later / %
Negril	40	6
Chalet Caribe	80	20
Montego Bay	48	6
Rio Bueno	64	4
Discovery Bay	62	2
Pear Tree Bottom	70	6
Port Maria	42	2
Port Antonio	54	4
Port Royal Cays	22	4

Table 5.4: The decline in Caribbean reefs over a 20-year period.

a Calculate the drop in the mean coral cover of all the listed reefs.
   Step 1: Calculate the mean initial coral cover.
   Step 2: Calculate the mean coral cover 20 years later.
   Step 3: Calculate the drop in mean coral cover.

b Using the equation below, calculate the percentage change in coral cover for each of the reefs.

$$\text{Percent change} = \frac{\text{new} - \text{original}}{\text{original}} \times 100$$

3 Compare the biodiversity found at Reef A and Reef B in Table 5.5.
   Remember:
   - A comparison can list similarities as well as differences, so identify which taxa are the same.
   - It is always a good idea to support a statement with data from the table or graph being studied: for example, 'Reef A has 900 fish species.' When comparing two sets of data, it is best practice to do a calculation of the difference: for example, 'Reef A has 405 more fish species than Reef B.'

Taxon	Number of species found	
	Reef A	Reef B
fish	900	495
mammals	27	0
reptiles	15	15
crustaceans	143	46
mollusks	450	246
echinoderms	113	30
Total	1648	832

Table 5.5: Biodiversity at Reef A and Reef B.

Step 1: State which reef has a greater total number of species. Calculate the difference between the two reefs in the total number of species.

Step 2: State the differences in the number of species found between each taxon.

> **TIP**
>
> Make sure that it is clear which reef you are referring to. For example, do not simply say: 'It is rich in fish'. Instead, link your answer to both reefs: 'Reef A has more fish than Reef B' or 'Only Reef A has mammals'.

4 The world's largest reef is the Great Barrier Reef in Eastern Australia. Extensive aerial surveys and dives have revealed that 93% has been devastated by **coral bleaching**.

   a With reference to Figure 5.2, describe how marine scientists can study the biodiversity of a coral reef.

> **TIP**
>
> The context of the question is an underwater coral reef rather than the littoral rocky and sandy shores that you have already studied. Make sure that your answer specifically links to this figure and habitat.

Figure 5.2: Underwater population sampling of a coral reef.

   b Over the last few decades, the Great Barrier Reef has come under stress due to urbanisation. Suggest how this could cause a decrease in coral abundance.

   c Explain how a reduction in the coral populations could lead to a reduction in reef biodiversity.

   d Global warming has resulted in areas of temperate coral being replaced with tropical coral species. Discuss why an increase in temperature has resulted in this community change.

   e There is significant genetic biodiversity within coral species. Suggest how this knowledge may be useful when considering which variants to introduce to reefs in areas suffering from global warming.

5 a Abandoned boats, tanks and trains can also be used as **artificial reefs**. State one positive and one negative effect of using these recycled materials.

   b Artificial reefs can be made of porous as opposed to solid concrete. Predict how this could increase the biodiversity of a coral reef.

> **KEY WORDS**
>
> **coral bleaching:** whitening of coral that results from the loss of a coral's symbiotic zooxanthellae
>
> **artificial reef:** an underwater structure built by humans to mimic the characteristics of a natural reef

c   Figure 5.3 shows a design of artificial reef that uses both timber and concrete as construction material. State a problem of using timber in these designs.

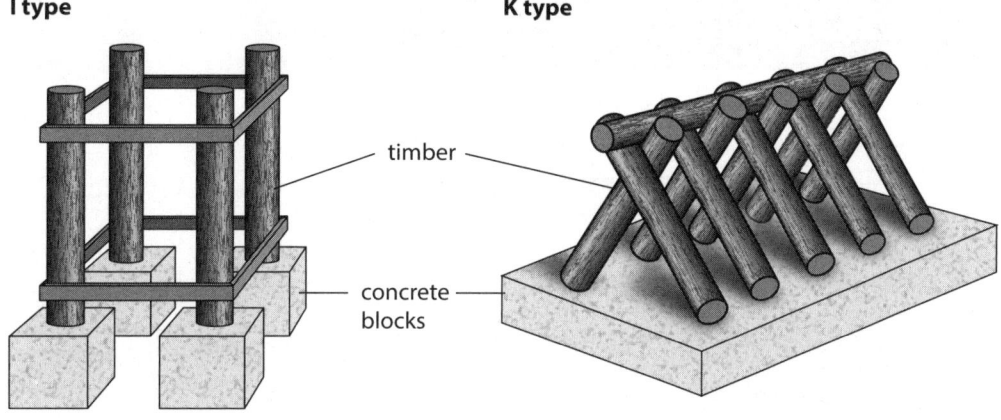

Figure 5.3: Artificial reef designs.

d   Suggest a potential benefit for each type of artificial reef.

Exercises 5.3 Sandy and rocky shores

Measurements can be incredibly large in marine science, but they can also be incredibly small. This exercise will increase your confidence in converting between SI units of length.

1   Another way to express small numbers clearly is to change the SI units that we are using.

   $1\,m = 1000\,mm$ millimetre
   $1\,mm = 1000\,\mu m$ micrometre
   $1\,\mu m = 1000\,nm$ nanometre

   Convert the units in these measurements:

   a   1 m into μm
   b   8 mm into μm
   c   600 μm into mm
   d   0.2 mm into nm
   e   50 nm into pm.

2   Plankton life that is washed up on a sandy shore can be defined by its relative size. Convert each of the maximum sizes of the plankton in Table 5.6 into millimetres.

Plankton type	Maximum size
megaplankton	200 cm
macroplankton	20 cm
mesoplankton	20 mm
microplankton	200 μm
nanoplankton	20 μm

Table 5.6: Maximum sizes of plankton.

3 Sand particles range in size from 0.0625 mm to 2 mm. The **Wentworth scale** can be used to describe different types of sand based on their grain size.

Wentworth scale	Max grain size / mm	Max grain size / μm	Permeability / darcy
very course sand	2		475
course sand	1		238
medium sand	0.5		119
fine sand	0.25		15
very fine sand	0.125		7

Table 5.7: Maximum grain size and permeability of different types of sand.

> **KEY WORDS**
>
> **Wentworth scale:** a way to describe different types of sand based on their grain size
>
> **porous:** substrate with holes that allow for the passage of air and water
>
> **permeability:** how well water flows through a substrate

a Copy and complete Table 5.7 to describe maximum sand grain size in μm.
b Draw a line graph to show the effect of sand grain size in μm on permeability.
c Describe the correlation between sand grain size and permeability.
d Sand particles are both **porous** (high porosity) and have a high **permeability** for water.
   i   State the effect of these properties on sand's ability to retain water.
   ii State what effect this has for the organisms adapted to live in sandy shores.

5 Compare the environmental factors affecting the community of organisms living on rocky and sandy seashores.

Exercise 5.4 The mangrove forest

Salinity plays significant roles in regulating the growth and distribution of mangroves. Mangroves have evolved a variety of salt tolerance mechanisms. This exercise will help you understand the adaptive features that allow different mangrove species to thrive in estuarine niches around the world.

1 a Copy and complete Table 5.8 to describe and explain the conditions required for the formation of mangrove forests.

Condition	Describe	Explain
temperature		
substrate		
wave action		
tidal range		
salinity of water		

Table 5.8: Conditions required for the formation of mangrove forests.

b Table 5.9 shows the results of a field experiment to survey the abundance of three species of mangrove at two marine reserves.

| Location | Tree cover / hectares | | |
	White mangrove	Black mangrove	Red mangrove
Marine Reserve A	40	290	170
Marine Reserve B	10	90	300

Table 5.9: Abundance of three species of mangrove at two marine reserves.

i   Plot a bar chart for the data in Table 5.9.
   ii   What percentage of mangrove forests at Marine Reserve A are red mangroves? Show your working.
2 Figure 5.4 shows the different root types found in black and red mangroves.

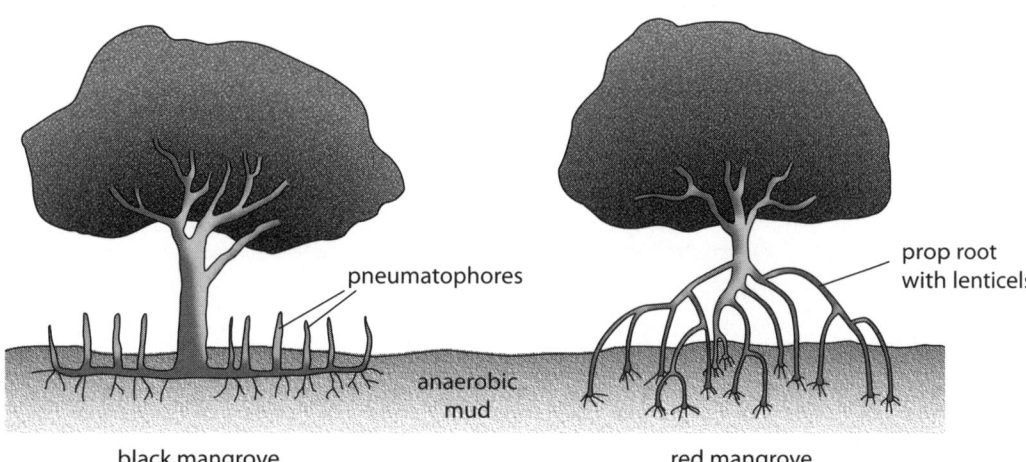

Figure 5.4: Root types found in black and red mangroves.

   a   Predict how the roots of black mangroves may deal with the lack of oxygen in the muddy substrate.
   b   In contrast to red and black mangroves, white mangroves live in coastal habitats above the high-tide mark. Suggest why white mangroves do not need either pneumatophores or lenticels.
3 Another environmental challenge faced by all aquatic plants is the loss or gain of water due to osmosis. **Osmosis** is when water enters or leaves plant cells. Water moves from an area of low salt or solute concentration to an area of higher salt or solute concentration.
   a   Suggest why many freshwater plants do not compete with mangroves as they cannot survive in estuarine conditions.
   b   The roots of some aquatic plants are salt accumulators. Predict why this prevents them dehydrating in salty conditions.
   c   Explain how the leaves of mangroves are adapted to reduce water loss.
   d   Discuss a benefit of the stems of some mangrove species being green.
   e   Salt tolerant plants (halophytes) can be either facultative or obligate.
       Facultative halophytes can grow in saline and non-saline habitats. **Obligate halophytes** can only survive in a saline environment. Justify why each definition can be used to describe mangrove forests.
4 Salt can inhibit the germination of mangrove seeds. Explain how **viviparous reproduction** by mangroves can increase the chance of germination.

> **TIP**
>
> You have studied red mangroves and will be required to apply this knowledge to the new context of other mangrove species.

> **KEY WORDS**
>
> **osmosis:** the movement of water from a higher water potential to a lower water potential across a selectively permeable membrane
>
> **facultative halophytes:** plants that can grow in saline and non-saline habitats
>
> **obligate halophytes:** plants that can only survive in a saline environment
>
> **viviparous reproduction:** [plants] a reproductive strategy where the seed develops into a young plant while still attached to the parent plant

EXAM-STYLE QUESTIONS

1 **a** Figure 5.5 shows the zones found in the open ocean. For each zone A–E, state the name and describe it in terms of light penetration. **[5]**

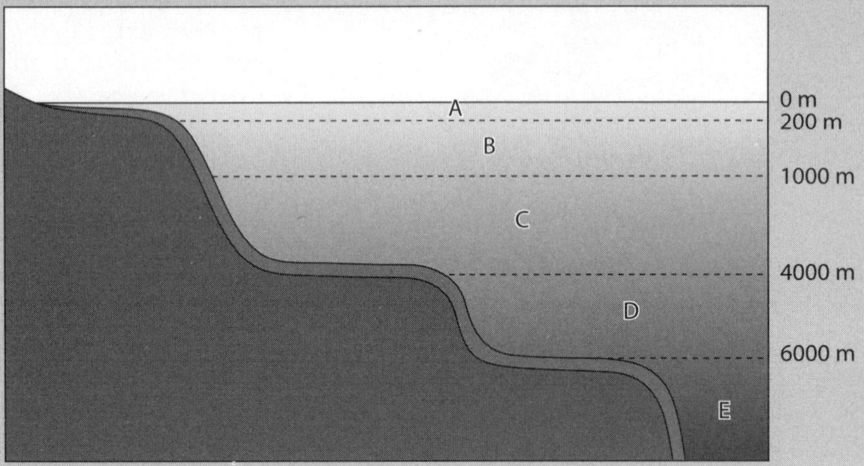

Figure 5.5: Open ocean zones

b Figure 5.6 shows the causes of changes in atmospheric CO_2 from 1870 to 2016. Calculate the total effect of fossil fuels on atmospheric CO_2. Show your working. **[1]**

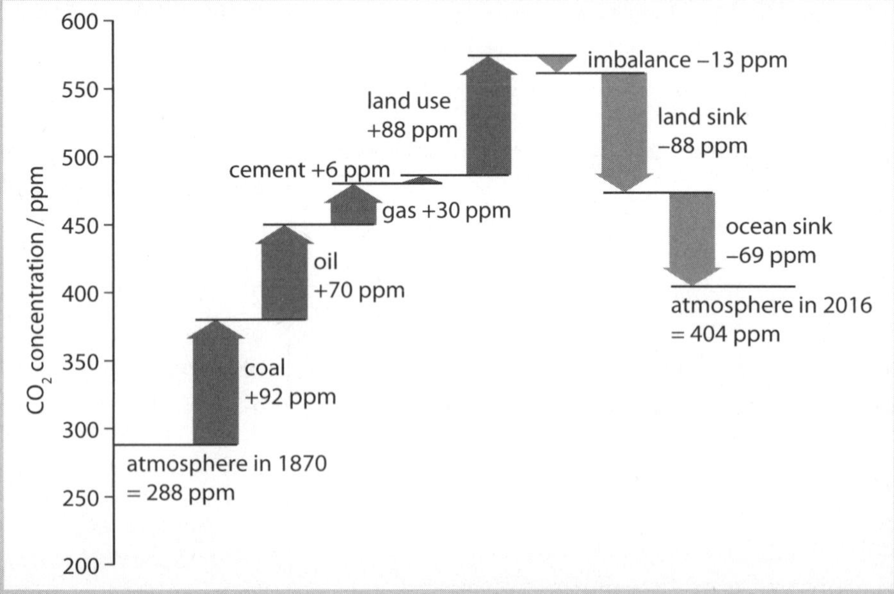

Figure 5.6: Causes of changes in atmospheric CO2 from 1870 to 2016

c Explain why oceans are described as carbon sinks rather than carbon sources. **[2]**

d Describe how oceans act as a carbon sink. **[3]**

CONTINUED

e In addition to being carbon sinks, state one other key interaction of oceans with the atmosphere. [1]

[Total: 12]

2 a Explain why coral reefs are most commonly found within 30°N and 30°S of the equator. [2]

b State the name of the products, labelled A and B, that pass between the zooxanthella and the coral polyp in Figure 5.7. [2]

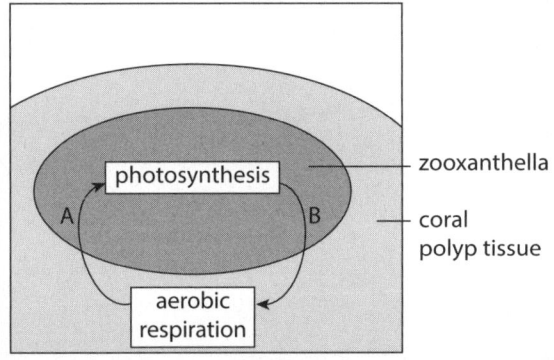

Figure 5.7: Mutulaistic relationship between zooxanthella and the coral polyp

c State which physical factor is most closely linked with the mutualistic relationship coral has with zooxanthellae. Support your answer with evidence. [2]

d Figure 5.8 shows the zonation of hard and soft coral.

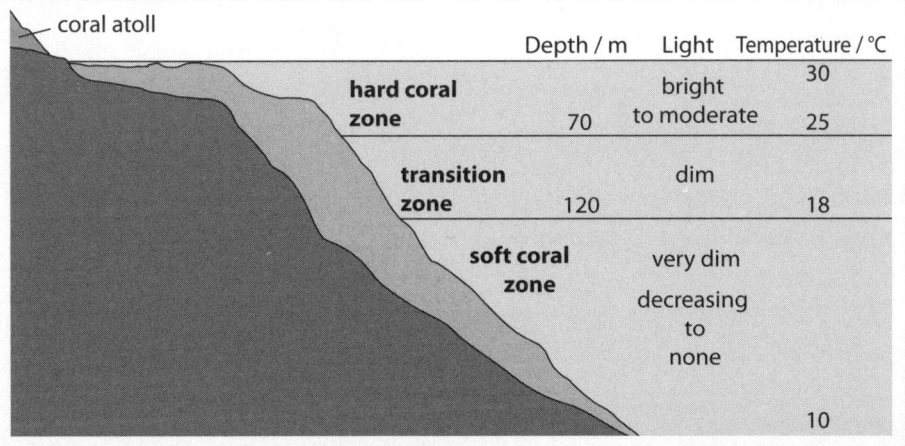

Figure 5.8: Coral reef zonation

Suggest how and why each of the following abiotic factors could affect the growth of the two types of coral.

   i light [2]
   ii temperature. [2]

[Total: 10]

CONTINUED

3
a Compare and **contrast** the lagoon structure in fringing reefs, barrier reefs, patch reefs and atolls. [4]
b Describe the major factors leading to reef erosion. [2]
c Explain the role of artificial reefs in the preservation of shorelines. [2]
d Suggest a reason why popular surfing locations might object to the installation of an artificial reef designed to protect the shoreline. [1]

[Total: 9]

4 a Satellite remote sensing now allows scientists to monitor global changes in the area covered by mangrove forests. Figure 5.9 shows the tree cover loss of mangroves from 2006 to 2009. Describe the difference in tree cover loss of mangroves over this period. [3]

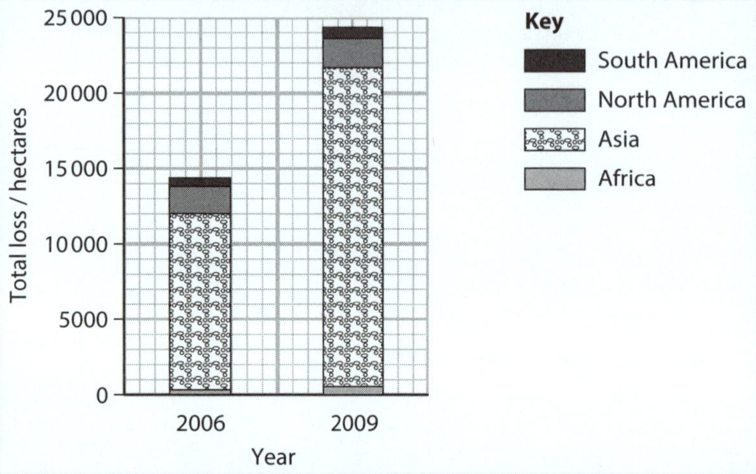

Figure 5.9: Mangrove tree cover loss from 2006 to 2009

b State *two* benefits of conserving mangrove habitat. [2]
c Explain how environmental changes may threaten mangrove forest. [8]

[Total: 13]

> **COMMAND WORD**
>
> **contrast:** identify / comment on differences

Chapter 6
Physiology of marine organisms

CHAPTER OUTLINE

The questions in this chapter cover the following topics:

- cell structures
- how to draw biological diagrams
- how to calculate magnifications
- how to draw graphs with two vertical axes
- the effect of salinity on rate of oxygen update and rate of drinking
- the movement of substances across membranes
- surface area: volume ratio
- gas exchange
- osmoregulation.

Exercises

Exercise 6.1 Drawing plan and high power diagrams

In Chapter 4, you will have practised your drawing skills and learnt that it is important to follow certain rules. Read through the rules again before attempting this exercise.

There are two main types of biological diagram that can be drawn of specimens: plan diagrams and high-power diagrams.

Plan diagrams show tissues and structures rather than individual cells and tend to have lower magnification. Do not draw individual cells when you draw these diagrams.

1   a   Figure 6.1 shows a light micrograph of an area of gill from a dogfish. Follow the steps below to draw a plan diagram of the area shown in Figure 6.1.

   Step 1: Get a sharp, hard pencil and a piece of blank A4 paper.
   Step 2: Draw the primary **lamellae**, secondary lamellae and capillaries. Figure 6.1 shows parts of six primary lamellae. The secondary lamellae are the projections on the sides of the primary lamellae. Inside the walls of the lamellae are capillaries.

> **TIP**
>
> Always draw what the specimen shows. Do not be tempted to draw idealised diagrams with structures that you think should be there.

> **KEY WORD**
>
> **lamellae:** addition branches of gill filaments that increase the surface area of the gills; they are sometimes called secondary lamellae

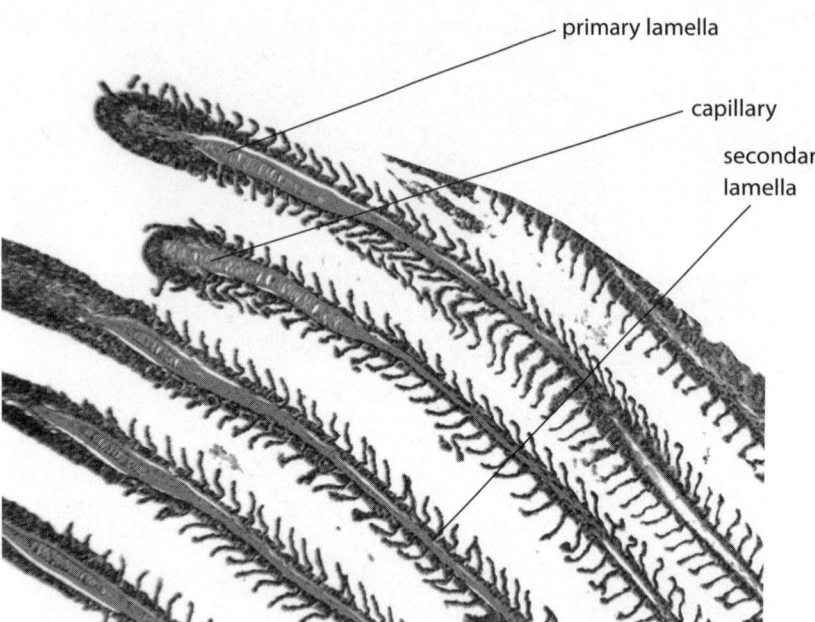

Figure 6.1: Light micrograph of an area of dogfish gill. Magnification: ×103 when printed at 10 centimetres high.

> **Step 3:** Label the structures by drawing straight label lines with a ruler. Use your pencil to print your labels and underline them.
>
> **Step 4:** Add a title to your diagram: 'Area of dogfish gill as shown by light microscope'.

   b Explain how the structure of these gills adapts them for rapid gas exchange.

2 High-power diagrams usually show cells or organelles as seen by high-power microscopy. You only need to draw a few cells to give a representative sample or, if appropriate, one cell with the organelles within it. The rules for drawing are the same as those for drawing plan diagrams.

   a Figure 6.2 is a light micrograph of red blood cells from a salmon. Fish red blood cells are different to human red blood cells because they have nuclei. Draw a labelled diagram to show the cells in Figure 6.2, with a ×2 magnification. Label a **nucleus**, a **cell surface membrane** and the cytoplasm of a cell. Add a title to the diagram.

   b Figure 6.3 shows a student's diagram of a cell from a green alga, as seen by a light microscope.

      List the mistakes that the student has made.

TIP

At A Level, you may have to magnify your drawing compared to the photograph. For example, a ×2 magnification would require you to do a drawing that is proportionally twice the size.

KEY WORDS

nucleus: membrane bound organelle that contains the genetic material of a cell

cell surface membrane: a biological membrane that separates the internal contents of a cell from its external environment

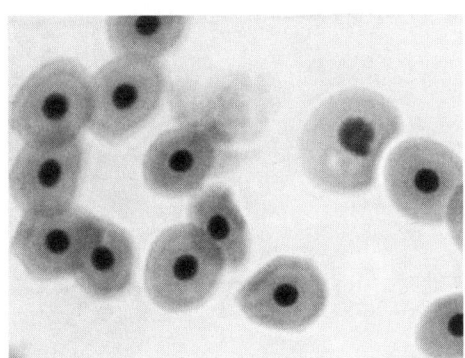

Figure 6.2: Red blood cells from salmon.

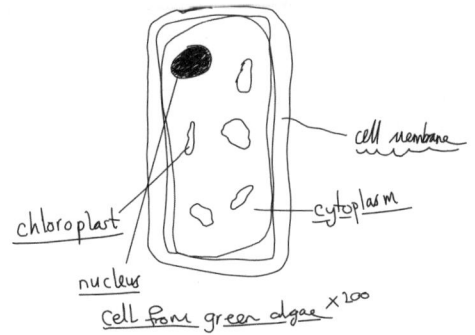

Figure 6.3: Student diagram of cell from green alga.

Exercise 6.2 Calculating magnifications

At A Level, you will need to be able to calculate magnifications of diagrams and the actual sizes of structures that are shown in images. This exercise will help you practise these calculations.

You are expected to know the formula for magnification. It may be best to learn the formula as a formula triangle so that it can be easily rearranged.

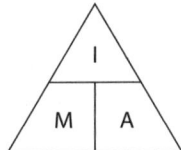

I – image size
M – magnification
A – actual size

Chapter 6 of the Coursebook gives full explanations of how to calculate magnifications and image sizes from photographs and diagrams with and without scale bars. Read through these pages before attempting the following calculations.

1   Calculate the magnifications of the following:

   a   a primary lamella of a fish gill with an image size of 25 mm and actual size of 5 mm

   b   a **mitochondrion** with an image length of 70 mm and an actual length of 5 mm

   c   a dolphin with an image size of 15 cm and an actual length of 1.8 m

   d   the magnification of the diatom shown in Figure 6.4 (actual width between A and B: 40 mm).

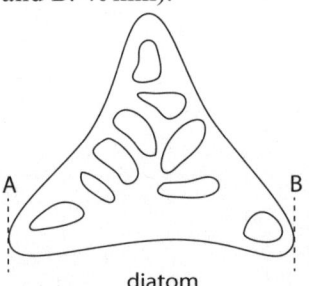

Figure 6.4: Diagram of a diatom.

TIP

Do not forget to make sure that you have the same units for image size and actual size. Always check your answer to make sure that it makes sense – if your chloroplast seems to measure 2 km in length, this suggests that you have made a mistake.

KEY WORD

mitochondrion: (plural: mitochondria) the organelle in eukaryotes in which aerobic respiration takes place

2 Calculate the actual sizes of the following:
   a   the length of the coral polyp (between A and B) shown in Figure 6.5
   b   the mean tentacle length of the coral polyp shown in Figure 6.5

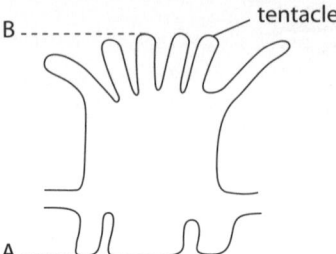

Figure 6.5: Diagram of a coral polyp, magnification ×2.5.

   c   the lengths of the head, midpiece and tail of the salmon sperm cell shown in Figure 6.6a.

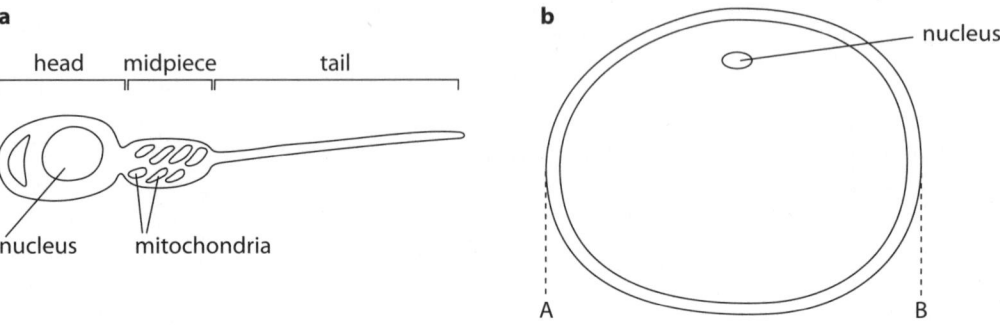

Figure 6.6: Diagram of salmon gametes (a) sperm, magnification ×1250, (b) ovum (egg), magnification ×15.

   d   the width of the salmon ovum (egg), between A and B, shown in Figure 6.6b.
   e   Suggest reasons for the differences in size between the salmon egg and sperm shown in Figure 6.6a.

4 Use the scale bars in the diagrams to calculate the following:
   a   the maximum length of the cell from the salmon intestine shown in Figure 6.7
   b   the maximum length of the nucleus in the cell from the salmon intestine shown in Figure 6.7
   c   the maximum width of the mitochondrion shown in Figure 6.8
   d   the mean length of the cristae inside the mitochondrion shown in Figure 6.8.

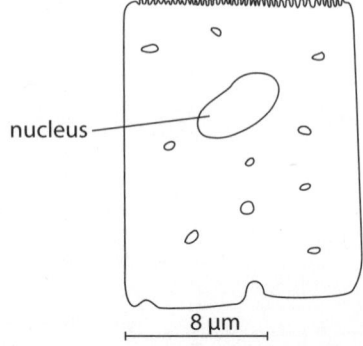

Figure 6.7: Diagram of a cell from salmon intestine.

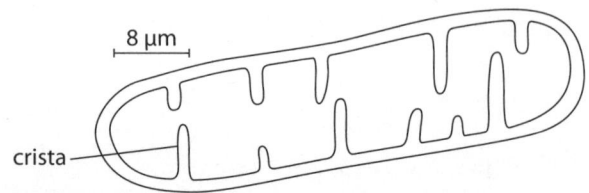

Figure 6.8: Diagram of a mitochondrion.

6 Physiology of marine organisms

Exercise 6.3 Matching organelles to their functions

For A Level marine science, you need to be able to recognise and know the functions of a range of cell organelles. This exercise will help you to summarise the structures and functions of the main cell organelles.

1 Copy and complete Table 6.1 to identify which of the organelles and structures are found in only plant cells, only animal cells or both types of cells.

Organelle name	Plant, animal or both	Structure	Function
cell surface membrane			
cell wall			
nucleus			
Golgi body			
mitochondria			
chloroplast			
large permanent vacuole			
ribosomes			
rough endoplasmic reticulum			
smooth endoplasmic reticulum			

Table 6.1: Comparing organelle structures and functions.

2 Look at the diagrams of different organelles and structures in Figure 6.9. Copy out the correct structures into your table.

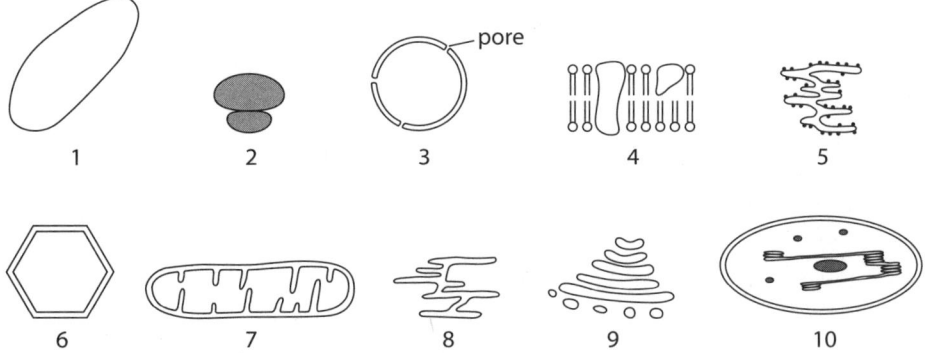

Figure 6.9: Diagrams of different cell organelles and structures.

KEY WORDS

cell wall: a layer that surrounds some types of cells and gives strength and support; plant cell walls are made of cellulose

Golgi body: cell organelle that modifies proteins

chloroplast: the photosynthetic organelle in eukaryotes

large permanent vacuole: a membrane bound organelle that is present in all plant and fungal cells; it contains cell sap

ribosomes: small organelles that are involved in the synthesis of proteins

rough endoplasmic reticulum (rER): a network of flattened membranous sacs, covered with ribosomes, that runs through the cytoplasm of a cell; proteins are synthesised in it before being transported to the Golgi body in vesicles

smooth endoplasmic reticulum (sER): a network of flattened membranous sacs that is found in the cytoplasm of cells; it is distinguished from the rough endoplasmic reticulum by its lack of ribosomes; its main function is the synthesis of lipids

3   The following box has a list of the functions of each organelle and structure. Copy out the correct function for each organelle and structure into your table.

> **TIP**
>
> Always check your plots carefully and make sure they are accurate to half a square or less.

Cell or organelle and structure functions

- Site of the synthesis of secreted proteins. It consists of a network of membranes that are found throughout the cell.
- Contains nucleic acids in the form of chromatin. It controls the activities of the cell. Has a double membrane surrounding it.
- Small organelles that synthesise proteins. They are either free in the cytoplasm or attached to internal cell membranes
- A network of membranes within the cytoplasm where some lipids and steroid hormones are made.
- Consists of an outer membrane and an inner membrane that is folded to increase its surface area. Site of aerobic respiration.
- Structure that is made of cellulose and is found around plant cells. Prevents cells bursting and provides support.
- Structure that is made up of a phospholipid bilayer with proteins inside it. It surrounds the cytoplasm and controls the passage of substances in and out of the cell.
- Structure that is made of a series of membrane cisternae that are stacked up together. It modifies and packages proteins before they are secreted.
- A large sac-like organelle surrounded by a membrane called a tonoplast. It contains cell sap and stores water and various solutes.
- An organelle that is surrounded by a double membrane. It contains a membrane network containing thylakoids which have pigments such as chlorophyll. Its function is photosynthesis.

Exercise 6.4 Drawing graphs with two y-axes

At A Level you are expected to be able to draw graphs with two dependent variables that may have very different magnitudes or units of measurement. This can mean that you need to have two separate vertical (y) axes and the same horizontal (x) axis. This exercise will help you to develop your graph drawing skills.

1   a   An investigation into the effect of temperature on the rate of gill operculum opening of a grouper was carried out. The effect of temperature on the concentration of oxygen in the water was also investigated. The results are shown in Table 6.2.

Follow the steps to draw a graph to display the data.

Temperature / °C	Rate of operculum opening / openings min^{-1}	Dissolved oxygen concentration / mg dm^{-3}
5	15.5	11.0
10	63.6	9.0
15	89.4	8.0
20	113.6	7.0
25	113.6	6.2
28	137.6	5.6

Table 6.2: The effect of temperature on the rate of operculum opening of a grouper and the dissolved oxygen concentration of the water.

Step 1: Draw a left-hand, vertical axis, and label it, 'rate of operculum opening / openings min^{-1}'. Choose a sensible linear scale with even increments.

Step 2: Draw a horizontal axis and label it, 'water temperature / °C'. Choose a sensible linear scale with even increments.

Step 3: Plot the data points for the rate of operculum opening. Join the points with ruled, straight lines. Do not go beyond the first and last points. Add a key for the line.

Step 4: Draw a second, vertical axis on the right-hand side of the graph. Label this axis, 'dissolved oxygen concentration / mg dm^{-3}'.

Step 5: Using the right-hand axis, plot the points for the oxygen concentration. Join the points with ruled, straight lines that are dashed. Add a key for the line.

    b Comment on the effects of temperature on the rate of operculum opening and dissolved oxygen concentration.

    c Suggest explanations for the effect of temperature on the rate of gill operculum opening.

2 The changes in population density of a species of mussel with distance from an estuary was determined. The salinity of the water from the estuary was also measured. The results are shown in Table 6.3.

Distance from river mouth into ocean / m	Mussel population density / mean number m^{-2}	Salinity of water / ppt
0	4	2
5	5	10
10	17	15
15	19	25
29	11	30
32	9	33
35	2	35

Table 6.3: Mussel population density and water salinity at different distances from river estuary.

a Draw a line graph with two *y*-axes to display the results in Table 6.3.
b Comment on the change in population density of the mussel species with distance from the estuary.
c Mussels are osmoconformers. Explain what is meant by the term *osmoconformer* and use your knowledge to explain the changes in population density of the mussels with distance from the estuary.
d Suggest how the population density of the mussels could have been determined.

EXAM-STYLE QUESTIONS

1 The effect of salinity on the rate of oxygen uptake, and the rate of water drinking by a euryhaline species of trout was investigated.

Five trout were placed into a tank of water with water of a salinity of 0 ppt. The volume of oxygen removed from the water was measured after 2 hours and the rate of oxygen consumption calculated as the mass of oxygen consumed per hour per kilogram of fish. The mean rate of drinking was also determined as the volume of water consumed per hour per kilogram of fish.

The results are shown in Table 6.4.

Salinity / ppt	Rate of oxygen consumption / mg hr^{-1} kg^{-1}	Rate of drinking / cm^3 hr^{-1} kg^{-1}
0	550	0.05
5	490	0.07
10	420	0.06
15	310	2.1
20	360	4.1
25	450	8.5
30	480	9.1
35	550	9.6
40	620	12.5

Table 6.4: The effect of salinity on the rate of oxygen consumption and rate of drinking of trout

a i State what is meant by the term *euryhaline*. [1]
   ii Give *two* variables that should have been kept constant in the investigation. [2]
b i Draw a line graph to show the change in rate of oxygen consumption and rate of drinking with increasing salinity. [6]
   ii Describe the effect of increasing salinity on the rate of oxygen consumption. [2]
   iii Use your knowledge of osmoregulation to explain the effect of increasing salinity on the rate of oxygen consumption and rate of drinking. [5]

[Total: 16]

CONTINUED

2 Figure 6.10 shows a diagram of an area of cell surface membrane of a cell in a red alga.

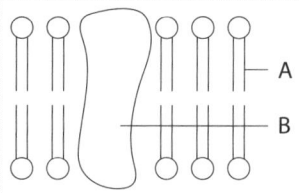

Figure 6.10: Diagram of an area of cell surface membrane

- **a i** Name molecules A and B. [2]
- **ii** Explain how the structure of the cell surface membrane controls the transport of substances into and out of cells. [5]
- **b i** Red algae cells contain chloroplasts. Draw a labelled diagram of a chloroplast. [4]
- **ii** Explain how chloroplasts are adapted to maximise photosynthesis. [4]

[Total: 15]

3 a i Compare and contrast *facilitated diffusion* and *active transport*. [4]
- **ii** Figure 6.11 shows the water potentials in three cells in a leaf of a seagrass plant. Draw arrows on the diagram to show the movement of water between the cells by osmosis. [1]

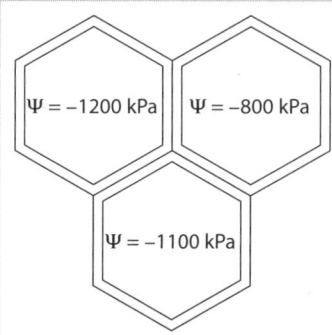

Figure 6.11: Water potentials in three cells in the leaf of a seagrass plant

b Figure 6.12 shows an osmometer. An osmometer is used to compare the water potentials of solutions. It consists of a bag made of selectively permeable Visking tubing. The bag is filled with a solution of 30 ppt salinity. The bag is attached to a glass tube through which the solution can move up and down if its volume increases or decreases.

CONTINUED

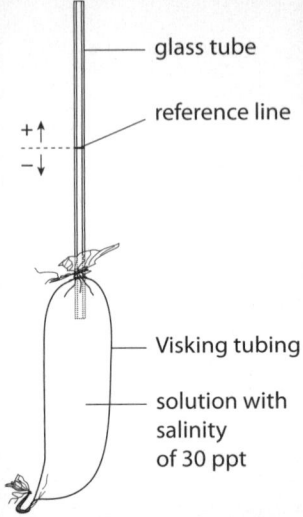

Figure 6.12: Diagram of an osmometer

An experiment was carried out into the effect of lowering an osmometer into water of different salinities.

The Visking tubing was placed into solutions of different salinities for one hour each. The distance that the solution moved compared to the reference line was measured in millimetres. The results are shown in Figure 6.13.

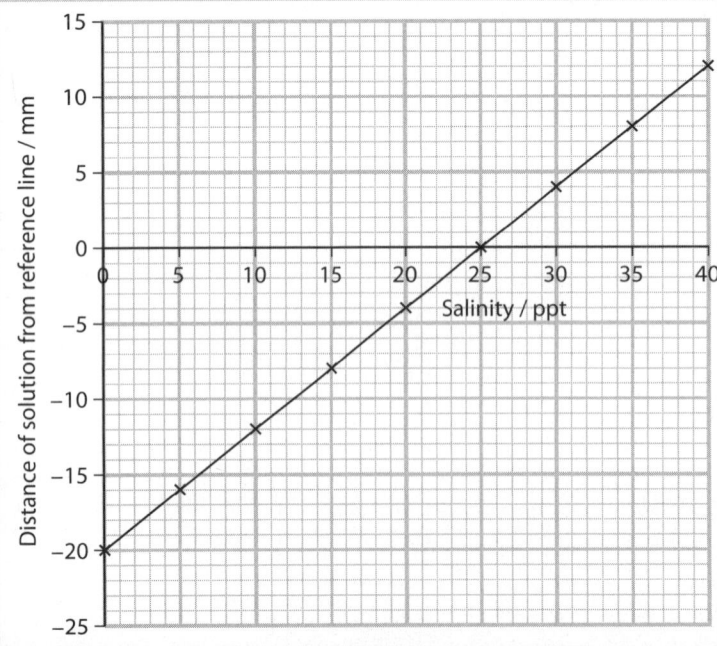

Figure 6.13: Results of the effect of salinity on the movement of solution in an osmometer

CONTINUED

    i Explain why the solution moved up and down from the reference line when the bag was placed into solutions of different salinity. [5]

    ii Use the graph to predict the salinity of a sample of seawater that caused the solution to move 7 mm from the reference line. Show your working on the graph. [2]

    iii Explain the problems that a stenohaline marine fish would have if placed into a salinity of 5 ppt. [3]

    [Total: 15]

4 Table 6.5 shows the surface area, volume and surface area : volume ratio of spheres of different radii.

Radius / mm	Surface area / mm²	Volume / mm⁻³	Surface area : volume ratio / mm⁻¹
1	12.56	4.186666667	3
10	1256	4186.666667	0.3
100	125 600	4 186 666.667	0.03
1000	12 560 000		

Table 6.5: The surface area, volume and surface area : volume ratio of spheres with different radii

  a **i** Calculate the volume of a sphere with radius of 1000 mm. [1]

$$V = \frac{4}{3}\pi r^3$$

    ii Use your answer to part **a i**, and Table 6.5, to calculate the surface area : volume ratio for a sphere with radius 1000 mm. [1]

    iii **Discuss** how the information in Table 6.5 explains the need for specialised gas exchange organs and a transport system in larger marine organisms. [4]

  b Diffusion can be investigated by placing agar jelly that contains sodium hydroxide and an indicator called cresol red. The indicator is red when in alkaline conditions and turns yellow when neutral or acidic. Cubes of alkaline, red agar jelly can be placed into hydrochloric acid. As the hydrochloric acid diffuses into the agar, it neutralises the alkali and the blocks turn yellow. The time taken for the block to turn yellow is a measure of the speed at which the acid has diffused to the centre.

Cresol red is an irritant. Sodium hydroxide and hydrochloric acid are corrosive.

Plan an investigation into the effect of changing the surface area to volume ratio of red agar blocks on the rate of diffusion of hydrochloric acid.

You should give full practical details including methods to minimise risks and draw a results table. [10]

[Total: 16]

> **COMMAND WORD**
>
> **discuss:** write about issue(s) or topic(s) in depth in a structured way

CONTINUED

5 a Describe the method by which grouper perform gaseous exchange. **[7]**

b State what is meant by the following terms:

　i osmoconformer **[1]**

　ii stenohaline. **[1]**

c Outline the mechanism of osmoregulation in salmon when in the marine environment. **[6]**

[Total: 15]

Chapter 7
Energy

CHAPTER OUTLINE

The questions in this chapter cover the following topics:
- the light dependent and light independent stages of photosynthesis
- the roles of photosynthetic pigments in photosynthesis
- the effect of light wavelength on photosynthesis
- the concept of limiting factors in photosynthesis
- the roles of accessory pigments in brown and red algae
- the process of aerobic respiration
- the process and importance of chemosynthesis
- the use of standard deviation as a measure of variation
- the effect of temperature and salinity on the rate of oxygen consumption.

Exercises

Exercise 7.1 Comparing photosynthesis and respiration

You need to know the words and balanced chemical symbol equations for **photosynthesis** and **aerobic respiration**. You should also understand the relationship between these two processes. This exercise will help your understanding of both processes and the principal of limiting factors related to photosynthesis.

1 a Write down the balanced chemical equations for photosynthesis and aerobic respiration.
  b State the relationship between photosynthesis and aerobic respiration as shown by the equations.
  c Copy and complete Table 7.1 to state how reducing each of the factors would affect the rates of respiration and photosynthesis. Record them as 'no effect' or 'reduce rate.'

Factor	Aerobic respiration rate	Photosynthesis rate
reduced light intensity		
reduced temperature		
reduced carbon dioxide concentration		
reduced oxygen concentration		

Table 7.1: Factors that affect aerobic respiration and photosynthesis.

> **KEY WORDS**
>
> **photosynthesis:** the process of using light energy to synthesise glucose from carbon dioxide and water to produce chemical energy
>
> **aerobic respiration:** the release of energy from glucose or another organic substrate in the presence of oxygen; the waste products are carbon dioxide and water

2 Marine algae both photosynthesise and respire aerobically. A student set up an experiment to compare the effect of light intensities on the rates of respiration and photosynthesis. The student placed pieces of marine algae into 25 cm^3 of hydrogen carbonate indicator in three test tubes as shown in Figure 7.1. They also set up a fourth test tube with no algae. The test tubes were placed in front of a lamp and the change in colour of the hydrogen carbonate indicator solutions recorded after four hours.

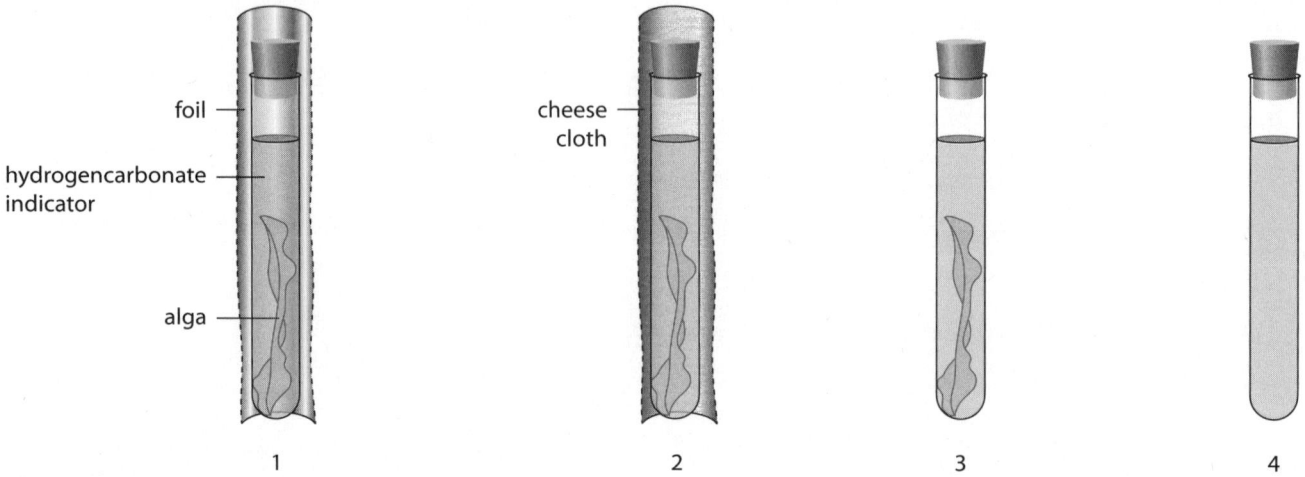

Figure 7.1: Experiment to compare effect of light intensity on the rates of respiration and photosynthesis.

Hydrogen carbonate indicator solution changes colour according to the concentration of dissolved carbon dioxide gas:

- in atmospheric (0.3%) carbon dioxide it is a red / orange colour
- in higher concentrations of carbon dioxide (>0.3%) it is yellow
- in lower concentrations of carbon dioxide (<0.3%) it is purple.

The changes in colour of the hydrogen carbonate indicator for each tube are shown in Table 7.2.

Tube number	Start colour	Colour after four hours
1 (foil)	orange	yellow
2 (cheesecloth)	orange	orange
3 (direct light)	orange	purple
4 (no algae)	orange	orange

Table 7.2: Colours of hydrogen carbonate indicators after four hours.

a Explain why the student included a tube with no algae (tube 4).

b Suggest other variables that the student should have controlled.

c Explain, in terms of the relative rates of photosynthesis and respiration, the colour changes of each of the tubes.

d The student suggested that when cheesecloth is placed in front of the test tube that the marine alga was unable to respire or photosynthesise. Explain whether the student was correct.

3 The rate of photosynthesis of a plant can be determined by measuring the rate of oxygen production. Figure 7.2 shows the effects of light intensity and temperature on the rate of oxygen production. If oxygen production is negative, it means that the plant has a net intake of oxygen.

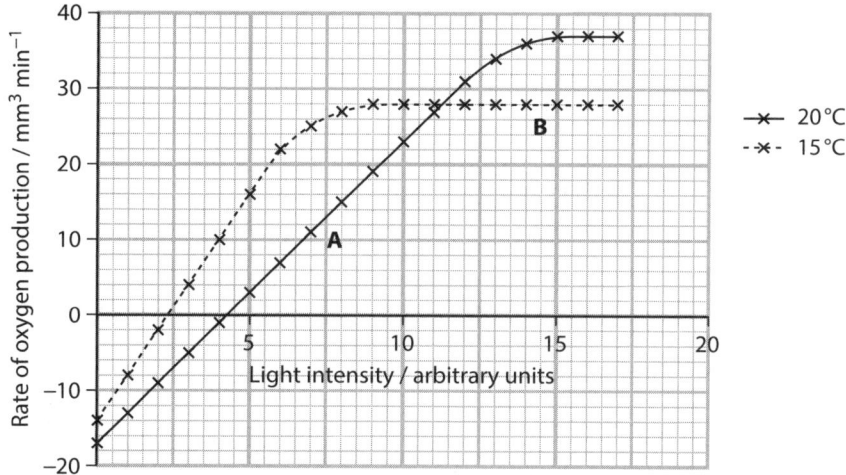

Figure 7.2: The effects of changing light intensity and temperature on the rate of oxygen production of marine algae.

TIP
In your explanations, you should try to explain all aspects of the data as fully as possible. Look for trends, turning points and possible outliers.

a State the light intensity at which the rates of respiration and photosynthesis are equal at 15 °C.

b State what factors are limiting the rates of photosynthesis at points A and B. Explain your reasoning.

c State the light intensity at which the rates of respiration and photosynthesis are equal when the temperature is 20 °C.

d Figure 7.2 presents complicated data. Discuss, as fully as you can, how both light intensity and temperature affect the rate of oxygen production.

e Use the information in the graph to explain: why marine algae cannot live in very deep water even though there is some light present; and why turbidity reduces algal growth more significantly in warmer water than in colder water.

Exercise 7.2 The effect of light intensity on the rate of light dependent stage, and calculating standard deviation

Photosynthesis has two distinct stages, the **light dependent stage** and the **light independent stage**. Light energy is harvested by **photosynthetic pigments** and converted into chemical energy in the forms of ATP and reduced NADP during the light dependent stage. This chemical energy is then used to synthesise carbohydrates in the light independent stage. This exercise will help you understand the light dependent stage and how to calculate standard deviation.

1 The chloroplasts from marine algae can be isolated in solution. If they are kept carefully, the light dependent stage can still occur. The photosynthetic pigments are oxidised by light and reduced NADP is produced. The rate

KEY WORDS

light dependent stage: the stage in photosynthesis whereby light energy is harvested; it occurs in the thylakoid membranes of chloroplasts and produces ATP and reduced NADP

light independent stage: the stage in photosynthesis whereby carbon dioxide is converted into glucose by the Calvin cycle; occurs in the stroma of chloroplasts

photosynthetic pigments: pigments such as chlorophyll that are used to absorb light during photosynthesis

of photosynthesis can be determined by adding a chemical called DCPIP (2,6-dichlorophenolindophenol). As shown in Figure 7.3, DCPIP is blue in its oxidised state and goes colourless when reduced (gains an electron.) As the light dependent stage occurs, electrons are given to DCPIP so that is turns colourless.

DCPIP (oxidised) + electron ⟶ DCPIP (reduced)

BLUE   COLOURLESS

Figure 7.3: The colour change of DCPIP.

To investigate the light dependent stage, a student set up three test tubes:
- Tube 1: mixture of chloroplasts and DCPIP
- Tube 2: DCPIP alone
- Tube 3: mixture of chloroplasts and DCPIP. The tube was covered with foil.

All the test tubes were illuminated. It took 45 minutes for the DCPIP in tube 1 to turn from blue to colourless. The DCPIP did not change colour in tube 2 or tube 3.

a Use your knowledge of the light dependent stage of photosynthesis to suggest an explanation for why the DCPIP went colourless.

b State the products of the light dependent stage.

c State the functions of the products of the light dependent stage in the **Calvin cycle**.

d Explain why the student included tube 2 and tube 3.

2 Plan an investigation, using DCPIP, to investigate the effect of different light intensities on the rate of the light dependent stage.

You should:
- State the independent variable and suggest how you could change it.
- State the dependent variable. Explain how you will ensure validity and how you will measure it.
- State all other variables that need controlling. Explain why each variable must be controlled and how you will do it.
- Suggest how you will analyse your data – what graphs you will plot or statistical tests you could do.

Experiments must be repeated several times to calculate mean values. Sometimes it is useful to know how much the data spreads about a mean. There are two measures that you need to be aware of: the **range** and **standard deviation**.

- The range tells us the highest and lowest values.
- The standard deviation tells us the values within which either 68% or 95% of values lie. If the repeated data points are similar, the standard deviation will be low. If the repeated data points are very different, the standard deviation will be high. Standard deviation is usually more useful than range because it discounts outliers.

For example, if the student repeated the experiment with DCPIP and obtained the data shown in Table 7.3, you can calculate the range and standard deviation to see how much variation there is.

> **KEY WORDS**
>
> **Calvin cycle:** the series of reactions that occur during the light independent stage of photosynthesis; it converts carbon dioxide and other substances into glucose
>
> **range:** the maximum variation of a data set from the lowest to the highest value
>
> **standard deviation:** a measure of the spread of data about the mean value

3   Follow the steps to calculate the range and standard deviation of the data.

Experiment number	Time taken for DCPIP to change colour / min	$(x - \bar{x})$	$(x - \bar{x})^2$
1	45	45 − 43 = 2	$2^2 = 4$
2	47		
3	42		
4	49		
5	51		
6	39		
7	41		
8	37		
9	50		
10	32		
	mean ($\bar{x}$) = 43		$\Sigma(x - \bar{x})^2 =$

Table 7.3: Repeated times for DCPIP to change colour.

Step 1: Copy out Table 7.3.

Step 2: In the column labelled $(x - \bar{x})$, subtract the mean from each value. The first is done for you.

Step 3: In the column labelled $(x - \bar{x})^2$, square the number in the previous column. The first is done for you.

Step 4: Calculate $\Sigma(x - \bar{x})^2$. This means add up all the numbers in the column labelled $(x - \bar{x})^2$.

Step 5: The formula for standard deviation is:

$$\sigma = \sqrt{\frac{\Sigma(x - \bar{x})^2}{n - 1}}$$

where $\sigma =$ standard deviation

$\Sigma(x - \bar{x})^2 =$ the sum of all the $(x - \bar{x})^2$

$n =$ the number of terms.

Substitute the numbers into the formula to calculate the standard deviation to one decimal place.

Step 6: Calculate the interval between which 68% of values lie. This is the mean ± 1 standard deviation.

Record this as mean ± 1 standard deviation *and* as the values between which 68% lie:

lower value → higher value

Also, calculate the interval between which 95% of values lie. This is the mean ± 2 standard deviations.

Step 7: Write down the range in the form of $x \to y$, where x is the smallest value and y is the highest value.

4 A student investigated the effect of light colour on the rate of decolourisation of DCPIP. The results are shown in Table 7.4.

Trial number	Time taken to decolourise DCPIP / min		
	red light	yellow light	blue light
1	59	275	49
2	63	290	64
3	54	260	45
4	41	310	49
5	58	315	55
6	72	275	58
7	49	290	61
8	61	310	57
9	49	250	44
10	60	310	68

Table 7.4: Results of investigation into effect of light colour on rate of decolourisation of DCPIP.

a Calculate the mean times and standard deviations of the times taken to decolourise DCPIP for the following:

  i red light
  ii yellow light
  iii blue light.

  In each case, give the 68% and 95% of values for the times taken to decolourise DCPIP in each colour of light, in this form: mean ± standard deviation.

b Use your knowledge of photosynthetic pigments to explain the results.

Exercise 7.3 The relative contributions of chemosynthesis, filter feeding and decomposition on deep-sea food webs

In addition to your AS Level knowledge of food webs, at A Level, you need to understand the role of chemosynthesis in marine food webs. This exercise will help your understanding of the process of chemosynthesis.

1 The source of organic molecules for deep-sea food webs at three different locations was investigated. The locations were: an area of high hydrothermal vent activity (HA); an area of moderate hydrothermal vent activity (MA); and an area of seabed with no hydrothermal vent activity (NH). Table 7.5 shows the percentage contribution of organic material to deep-sea food webs due to:

  • chemosynthesis by bacteria (C)

  • suspended particles of organic material that have descended from surface waters and are extracted by filter feeding organisms (SF)

  • decomposition of organic material that has descended from surface waters (DO).

Location	Percentage of organic material from each source / %		
	Chemosynthesis (C)	Filter feeding (SF)	Decomposition (DO)
HA	30.6	22.6	46.8
MA	13.8	55.5	30.7
NH	0.4	0.8	98.8

Table 7.5: Comparison of origin of organic material for three different deep-sea food webs.

a   Explain what is meant by the term *chemosynthesis*.
b   Explain what is meant by the term *decomposition*.
c   Compare the energy source for the organic material produced by chemosynthesis compared to the organic material obtained from filter feeding and decomposition.

2 a   Draw a graph to display the data shown in Table 7.5.
b   Compare the percentage of organic material from each source for each of the three locations.
c   Suggest explanations for the different percentages of organic matter from each source at the three locations.

EXAM-STYLE QUESTIONS

1 a i   Complete the chemical equation for photosynthesis. [1]

$$\underline{\qquad} + 6H_2O \longrightarrow C_6H_{12}O_6 + \underline{\qquad}$$

  ii   Describe the light dependent stage of photosynthesis. [4]

b   The effect of light intensity on the rate of photosynthesis of seagrass was investigated. Five discs of seagrass leaves with a radius of 5 mm were taken and placed into a boiling tube filled with seawater, as shown in Figure 7.4.

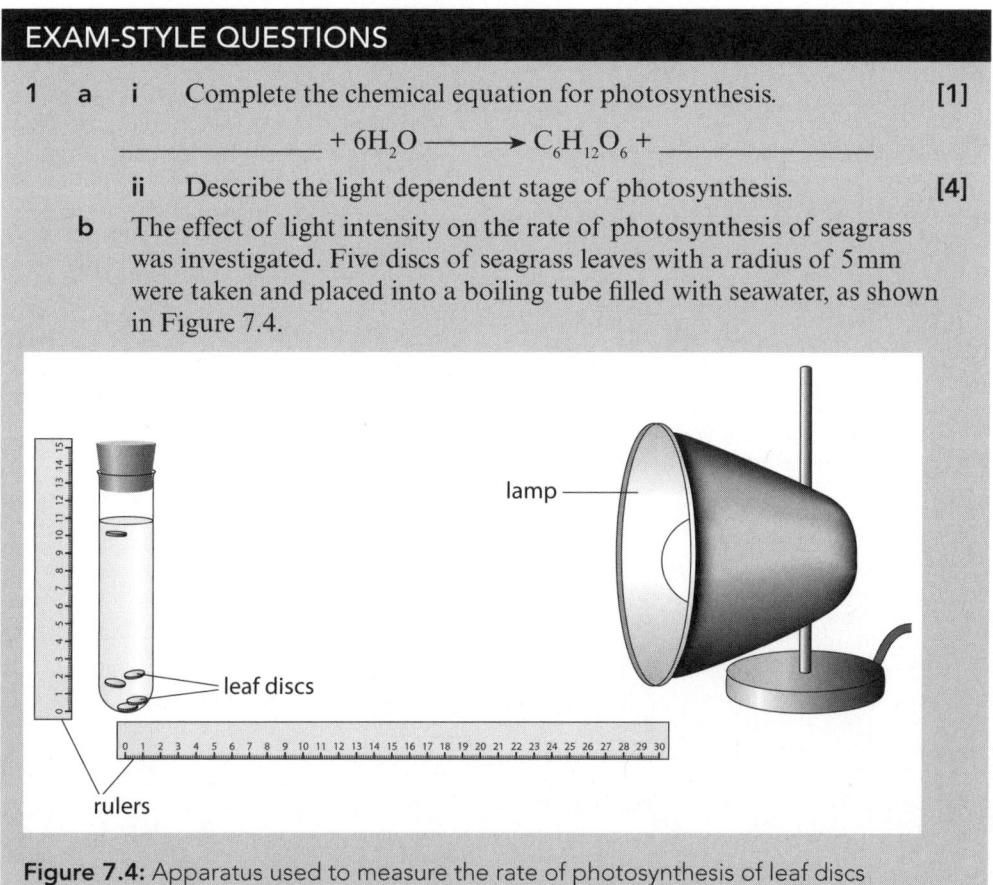

Figure 7.4: Apparatus used to measure the rate of photosynthesis of leaf discs

> **CONTINUED**

The boiling tube was illuminated, and the times taken for the discs to rise by 10 cm taken. The experiment was repeated with different light intensities. The results are shown in Table 7.6.

Light intensity / arbitrary units	Mean time taken for disc to rise 10 cm / s
0	290
1	570
2	180
3	120
4	100
5	80
6	70
7	
8	65
9	65

Table 7.6: Effect of light intensity on mean time taken for disc to rise 10 cm

   i The times taken for each of the five discs to rise 10 cm with a light intensity of 7 arbitrary units were:

      70 s 68 s 60 s 62 s 63 s

   Calculate the mean time taken for the discs to rise 10 cm.
Give your answer to two significant figures. [2]

   ii Draw a graph to show the effect of light intensity on the mean time taken for the discs to rise 10 cm.
Join the points with straight lines. [5]

   iii Discuss the effect of increasing light intensity on the mean rate taken for the discs to rise 10 cm. [5]

[Total: 17]

2 Chemosynthesis is a process that brings energy into ecosystems.

   a Compare and contrast the processes of chemosynthesis and photosynthesis. [4]

   b Scientists investigated the mean annual growth rate of *Riftia* at three hydrothermal vents. The temperature and the concentration of hydrogen sulfide in the water were also measured over the year. The mean results and their standard deviations are shown in Table 7.7.

CONTINUED

Vent	Mean annual growth rate / m yr⁻¹	Mean annual temperature / °C	Mean annual hydrogen sulfide concentration / mmole dm⁻³
1	27.2 ± 2.4	15.5 ± 5.6	225.3 ± 6.5
2	29.8 ± 1.8	17.6 ± 6.7	227.5 ± 5.8
3	52.3 ± 12.5	26.5 ± 15.4	495.6 ± 55.7

Table 7.7: Table showing the mean annual growth rates of *Riftia*, the mean annual temperatures and the mean annual hydrogen sulphide concentrations at three hydrothermal vents

    i State why the scientists calculated the standard deviations. [1]

    ii Compare the growth rates, mean annual temperatures and mean annual hydrogen sulfide concentrations of the three vents. [3]

    iii Discuss the mean annual growth rate shown by *Riftia* at vent number 3. [4]

[Total: 12]

3 a i State the location within chloroplasts of the light independent stage of photosynthesis. [1]

    ii State the name of the enzyme that is used in carbon dioxide fixation. [1]

    iii Name the products of the light dependent stage of photosynthesis that are used in the light independent stage. [2]

b A student investigated the effect of different light wavelengths on the rate of photosynthesis of an aquatic plant. Figure 7.5 shows the apparatus that the student used.

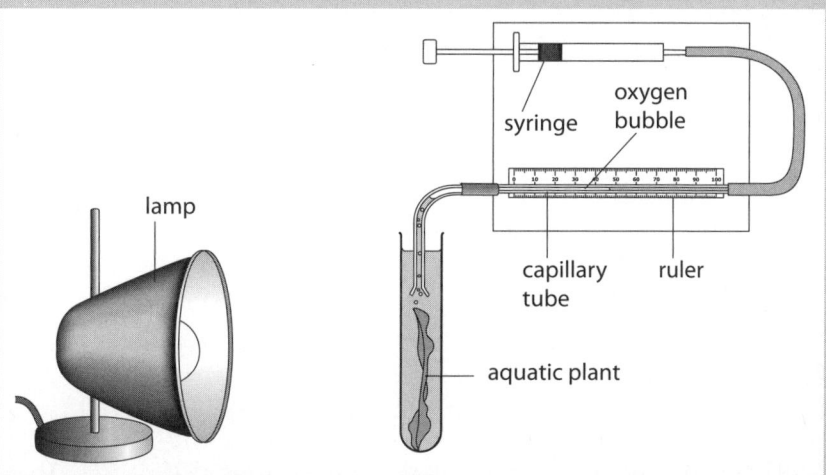

Figure 7.5: Apparatus used to investigate rate of photosynthesis of an aquatic plant

CONTINUED

The student illuminated the plant with light of wavelength 385 nm for 15 minutes and collected a bubble of oxygen gas in the capillary tube. The length of the bubble of oxygen, and the radius of the capillary tube, were measured. The method was repeated with light of different wavelengths. The results are shown in Table 7.8.

Wavelength of light / nm	Colour of light	Length of bubble of oxygen / mm	Rate of oxygen / mm^3 min^{-1}
385	violet	25	11.8
470	blue	27	12.7
540	green	2	0.94
580	yellow	9	
610	orange	27	12.7
690	red	28	13.2

Table 7.8: Table showing the effect of different light wavelengths on the rate of oxygen production by an aquatic plant

i   The radius of the capillary tube was 1.5 mm. Use the formula for the volume of a cylinder to calculate the rate of oxygen production in mm^3 min^{-1} with light wavelength of 580 nm. Give your answer to three significant figures. [3]

$$V = \pi r^2 d$$

ii   Use your knowledge of photosynthetic pigments to explain the results in Table 7.8. [3]

iii   Red light often does not penetrate below a depth of 6 m. Explain how brown and red algae are able to survive below this depth. [4]

[Total: 13]

CONTINUED

4 A student investigated the effect of increasing water temperature on the rate of aerobic respiration of mussels. They placed mussels in a tank of seawater at a temperature of 10 °C and measured the change in oxygen concentration with an oxygen meter over a one-hour period. They repeated the experiment at a range of temperatures. Their results are shown in Figure 7.6.

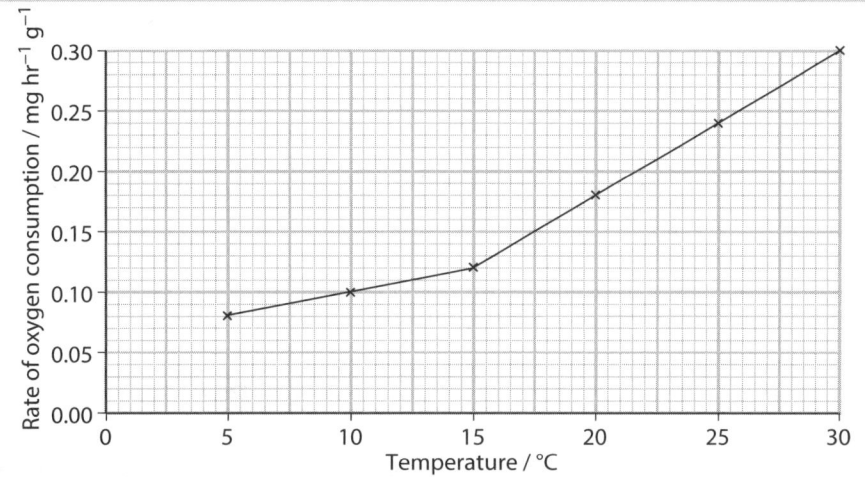

Figure 7.6: Graph to show the effect of temperature on the oxygen consumption of mussels

  a **i** Give the balanced chemical symbol equation for aerobic respiration. [2]

  ii Compare and contrast aerobic and anaerobic respiration in fish. [3]

  iii Describe the effect of temperature on the rate of oxygen consumption by the mussels. [2]

  b Plan an investigation into the effect of salinity on the respiration of mussels. You are provided with a stock solution of sea water with a salinity of 35 ppt, pure water, glass tanks, an oxygen meter and other standard laboratory equipment. Make sure to clearly state the independent, dependent and control variables, give full experimental details, and give a statistical test that you would use to analyse the results. [10]

  [Total: 17]

5 **a** Explain the roles of named accessory pigments in deep-sea marine algae. [4]

  b Describe the light independent stage of photosynthesis. [6]

  c Explain the ecological importance of marine primary producers. [5]

  [Total: 15]

Chapter 8
Fisheries for the future

CHAPTER OUTLINE

The questions in this chapter cover the following topics:

- sessile and non-sessile stages of the life cycles of marine animals
- the differences between simple and complex life cycles by a range of marine organisms
- the advantages and disadvantages of different reproductive strategies adopted by tuna, sharks and whales
- the impact of modern fishing technology, including sonar, purse seine fishing and benthic trawling
- the tools required for the sustainable exploitation of marine organisms
- how to generate ecological models for the sustainable exploitation of fish stocks in Antarctic waters
- the social, economic and environmental impacts of aquaculture.

Exercises

Exercise 8.1 Reproductive strategies

This exercise will build your confidence with understanding different reproductive strategies adopted by a range of marine organisms.

The two main reproductive strategies are termed **K-strategy** and **r-strategy**. K-selection is carried out by species that produce few 'expensive' offspring (in terms of invested energy and time); r-selection is carried out by species that produce many 'cheap' offspring.

1. **a** State the name of the type of life cycle (**complex life cycle** vs. **simple life cycle**) and the reproductive strategies for:

    i tuna

    ii whales

    iii sharks.

   b Copy and complete Table 8.1 to describe the differences in characteristics between r- and K-strategists.

Reproductive characteristic	r-strategists	K-strategists
type of fertilisation		
organism size	smaller	
lifespan		long lived
maturity rate	early onset	
reproduction frequency		more than once
number offspring produced		

KEY WORDS

K-strategy: producing few offspring but providing a large amount of parental investment

r-strategy: providing large numbers of offspring while providing little parental investment

complex life cycle: cycle in which the pre-reproductive and reproductive stages are very different and may change morphology, habitat, and diet

simple life cycle: cycle in which pre-reproductive and reproductive stages are very similar to one another

Reproductive characteristic	r-strategists	K-strategists
energy invested in each offspring		
size at birth	small	
amount of parental care		

Table 8.1: Characteristics of r- and K-strategists.

c Antarctic squid have short lifespans, mature in less than a year and produce large numbers of young. Antarctic cod have low metabolic rates, mature late in life and produce only a few large yolky eggs.

  i Describe the kind of reproductive strategy the best describes each of the two species.

  ii Explain which species is more likely to be in danger from overfishing.

d State the type of reproductive strategy adopted by human beings and justify your answer.

Exercise 8.2: Sustainable fisheries

This exercise will use the context of the Galápagos Islands to explore the tools required for the sustainable exploitation of a range of its marine organisms.

1 For many years the Galápagos Islands Marine Reserve (GMR) (Figure 8.1) has struggled to protect the waters within 40 nautical miles of the islands.

> **TIP**
>
> This is an open-ended question, so a range of valid points will be suitable. That said, make sure that your answer is specific to the context of the Galápagos Islands' marine environment.

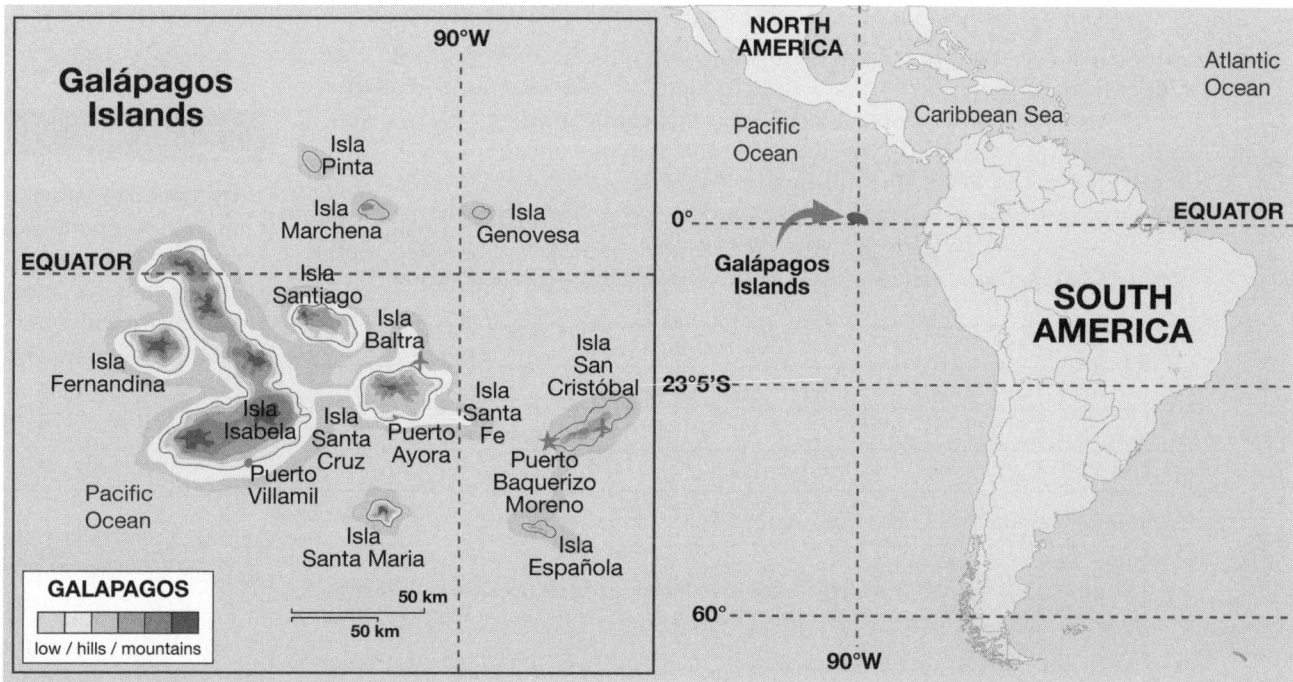

Figure 8.1: Map of Galápagos Islands.

Propose a potential solution for each of the problems listed below.

a   fishermen ignoring quotas

b   poor local law enforcement

c   growing international demand for shark fins

d   underfinancing of the GMR

e   patrol boats unable to catch poachers

f   difficulty of spotting illegal boats

g   poachers armed but GMR staff unarmed

2   a   Spiny lobsters are harvested in the GMR. Initially, there was no quota for lobster catches. As the number of fishermen rose in the 1990s, a lobster season was enforced but failed to protect the populations from collapse. Suggest reasons why this may have happened.

b   The Ecuadorian Government then took a number of actions to try to ensure that the lobster catch was sustainable. Discuss the potential positive and negative effects of each action listed below:

   i   a temporary ban

   ii   a reduced annual fishing season of four months

   iii a limited number of authorized fishing methods (for example, SCUBA and free diving)

   iv   minimum total length of 26 cm

   v   a ban on catching pregnant females

   vi   an annual quota based on the lobster population.

c   Table 8.2 shows the results from an investigation into the abundance of slipper lobsters at five fishing sites close to Santa Fe Island. At each site, the marine scientist also assessed the frequency of **benthic trawling** on a scale of 0–10, with 0 indicating no evidence of benthic trawling and 10 indicating the highest frequency of benthic trawling.

Site	Mean number of slipper lobsters per quadrat	Frequency of benthic trawling (1–10)
A	0	10
B	17	2
C	3	8
D	7	6
E	20	0

Table 8.2: Abundance of slipper lobsters at five fishing sites.

   i   Plot a line graph of the mean number of slipper lobsters (*x*-axis) against the frequency of benthic trawling (*y*-axis).

   ii   Describe the relationship between the mean number of lobsters and the frequency of benthic trawling.

   iii Suggest an explanation for the pattern in the data.

   iv   Suggest factors other than benthic trawling that may be responsible for the relationship pattern in (i).

TIP

An effective way to approach this question would be to create a table stating one advantage and one disadvantage of each measure.

KEY WORD

benthic trawling: a fishing method that drags a net along the seabed; wooden boards at the front of the net keep the net open and stir up the seabed, causing damage

TIP

Remember that correlation does not equate to causation. It is therefore incorrect to state that 'benthic fishing causes the death of slipper lobsters'.

Exercise 8.3: Sustainable fishing models

This exercise will build your understanding of the information required to generate sustainable models for fishing in Antarctic waters.

1   a   The Ross Sea is an important industrial fishing ground next to Antarctica. It has range of habitats for fish species. Define the location of the following habitats.

   - pelagic
   - demersal
   - benthic.

   b   Research vessels can use sonar to locate and predict the size and position in the water column of a shoal of fish. The distance to the shoal is calculated using the equation:

   $$\text{distance to shoal} = \text{speed of sound} \times \frac{1}{2} \text{time for sonar echo}$$

   i   Why is the time of a sonar echo divided by half?

   ii   If sound travels in water at 1440 m per second, and the sonar echo takes 10 seconds, how far away is the shoal?

   c   State another way that researchers could gather data on marine trophic relationships.

2   a   Table 8.3 includes data collected to calculate the **biomass** of fish species in the Mekong Delta.

Trophic group	Catch / tonne m^{-2} yr^{-1}	Fishing mortality
top predators	0.064	0.375
mackerel	0.017	1.000
benthic feeders	0.196	0.625
demersal fish	1.036	0.750
small pelagic fish	0.073	1.250
trashfish	0.585	1.700
shrimps	1.602	1.500
crabs	0.183	1.250
squids	0.616	1.250

Table 8.3: Mekong Delta data.

Biomass is calculated using the following equation:

$$\text{biomass} = \frac{\text{catch}}{\text{fishing mortality}}$$

   i   Calculate the biomass to 3 d.p. for each trophic group.

   ii   Calculate the mean biomass for the trophic groups.

3   Mathematical models can be used to calculate a theoretical basis for sustainable fishery management within the Antarctic. The most widely used ecosystem model for marine systems is called ECOPATH and it takes a snapshot of the structure of a food web at one time. An ECOPATH model investigating fishing in the East

> **TIP**
>
> Remember to show your working when doing any calculations. Good practice is to write an equation; show the data being processed; write the final answer to the appropriate number of decimal places and with the correct units.

> **KEY WORD**
>
> **biomass:** the mass of living material in an area; it can be measured as dry mass (without the water) or wet mass (with the water)

China Sea ecosystem showed high death rates for certain species in lower trophic levels that were not being commercially fished (for example, shrimps, crabs, anchovies and small fish).

a Discuss reasons why this might be so.

b What recommendations could scientists make to the fishing industry?

c Discuss why **trophic efficiency** may vary between trophic groups.

> **KEY WORD**
>
> **trophic efficiency:** the efficiency with which energy is transferred from one trophic level to the next

4 The Convention for the Conservation of Antarctic Marine Living Resources (CCAMLR) is the organisation that regulates fishing in the Antarctic. It uses a closely related ecosystem model called ECOSIM, which allows ecosystem managers to predict changes in the biomass of different trophic groups over time.

a What is the difference between the ECOPATH and ECOSIMS models?

b Describe the benefit for fishery managers of the ECOSIM model compared to the ECOPATH model.

c Suggest how ECOSIM modelling may be useful to Marine Conservationists.

Exercise 8.4 Aquaculture

This exercise will help you analyse and discuss the social, economic and environmental impacts of an aquaculture project in New Zealand.

1 The global aquaculture market was valued at US$170 billion in 2017. Create a pie chart to represent the value of each of the aquaculture habitats shown in Table 8.4.

Habitat	Value / US$ billion
freshwater	99
marine	46
brackish	25
Total	170

Table 8.4: Value of aquaculture habitats.

Step 1: Calculate the percentage of each habitat by dividing the habitat value by the total (170).

Step 2: Calculate the angle on the pie chart / 360 for each habitat.

Step 3: Draw the pie chart.

> **TIP**
>
> Give your maths a quick check by adding all the percentage and angle amounts. These should be approximately 100% and 360° but may differ very slightly due to rounding.

2 Brown trout (*Salmo trutta*) were introduced to New Zealand from Europe in the late 1860s and now occur throughout the South Island and lower North Island. They can spread by going out to sea and swimming up other rivers. The efficiency of New Zealand's brown trout farming has recently been investigated over an 8-year period. The results are shown in Table 8.5.

Year	2012	2013	2014	2015	2016	2017	2018	2019
Farming efficiency in arbitrary units / a.u.	34	19	67	18	42	34	41	29

Table 8.5: The efficiency of brown trout farming in New Zealand.

> **TIP**
>
> Ensure that the columns are drawn correctly; because this is a histogram of continuous data, the bars must touch. Only for a bar graph of non-continuous data should you draw spaces between the columns.

a   Calculate the mean annual farming efficiency to the nearest whole number.
b   Plot a histogram of the data in Table 8.5.
c   Despite being proposed twice, in 1970 and 2012, brown trout farming in New Zealand remains prohibited under the Conservation Act and Fisheries Act.
   i   Suggest reasons why the opening of commercial trout hatcheries is opposed by local salmon farmers.
   ii An alternative proposal to introducing trout farming is to encourage more mussel farming in New Zealand. Suggest why anglers and conservationists might encourage this option from an environmental perspective.

EXAM-STYLE QUESTIONS

1 a Figure 8.2 shows the life cycles of four marine species. Use the information given to:
      i   State a species with a complex life cycle.
      ii   Name a sessile larval stage.
      iii Name a planktonic stage. [3]

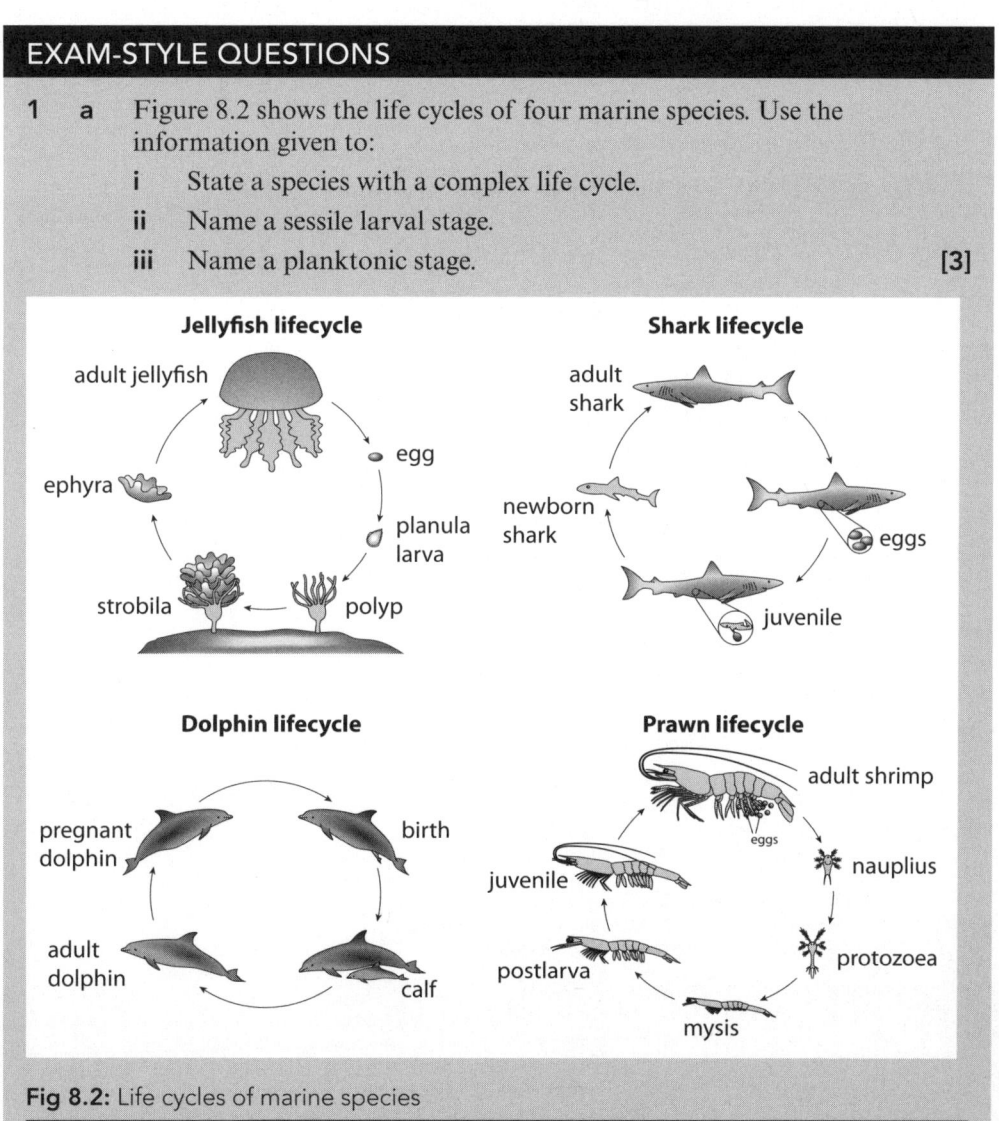

Fig 8.2: Life cycles of marine species

CONTINUED

b **Define** the difference between complex and simple life cycles. [2]

c Explain the advantages and disadvantage of a simple life cycle. [2]

d Like other whales, dolphins have internal fertilisation. State *one* disadvantage of this method of fertilisation. [1]

e Describe the different investments in the care of offspring shown by sharks. [3]

[Total: 11]

> **COMMAND WORD**
>
> **define:** give precise meaning

2 a The Galápagos Islands are a volcanic archipelago in the Pacific Ocean, 1000 kilometres to the west of Ecuador. One of the three main islands for fishing in the Galápagos is San Cristobal Island.

  i Between 1993 and 2000 there was an increase in the number of boats by 238%. If the number of boats registered in 1993 was 101, how many were there in 2000? [1]

  ii In 1997 there were 203 fishermen, while in 2000 there were 342. State the percentage increase in fishermen per year. [1]

b Initially fishing on the islands was artisanal fishing: traditional subsistence fishing by individual fishing households. Explain why this can be described as sustainable fishing. [1]

c The 1990s saw an explosion in numbers of boats from Ecuador who came to the Galápagos after the collapse of their own local fisheries. These new boats used tracking devices and longlining. Explain how each of these practices could damage the marine ecosystem. [2]

d Illegal commercial boats using purse seines took advantage of maritime law that allows a boat to enter prohibited waters of the Galápagos Marine Reserve (GMR) for up to 72 hours if it has a mechanical or medical emergency. Explain the possible impact of this on the GMR. [3]

e Some marine reserves also set a maximum catch size. State how this could help protect a recovering population. [1]

f Suggest how the marine scientific research by the Charles Darwin Foundation and the Galápagos National Park Service might help support the Galápagos Marine Reserve. [1]

[Total: 10]

3 a The US$4.04 billion global oyster farming industry has, in recent years, been ravaged by Pacific Oyster Mortality Syndrome. This syndrome is caused by a herpes virus that does not harm humans.

  i Pacific Oyster Mortality Syndrome has reduced the crop of Pacific cupped oysters in France from 110 800 tonnes in 2007 to 82 000 tonnes in 2012. Calculate the mean annual reduction in oyster production. Show your working. [2]

  ii The herpes virus starts killing oysters when the ocean temperature exceeds 16 °C. Suggest how human activity may be linked to this disease. [1]

CONTINUED

b Salmon poisoning caused by the bacterium *Nanophyetus salmincola* can infect cats that eat raw salmon. The bacterial pathogen is contained in a parasitic flatworm called a salmon fluke. This fluke has a complex life cycle that relies on three hosts: snails, salmon and cats. The fluke attacks a cat's lymph nodes and can result in death within a fortnight of infection.

  i Analysis of cats that ate contaminated raw salmon showed infection in 159 out of 305 dead cats and in 98 out of 257 live dogs tested. Calculate the percentage infected in the total sample of cats tested. Show your working. [2]

  ii Explain how salmon could be processed to ensure that it could be safely used in cat food. [3]

[Total: 8]

4 a We consume 5 600 000 tonnes of tuna globally per annum. Only 0.1% is currently farmed as opposed being caught in the wild. Estimate the amount of tuna farmed annually. Show your work. [1]

b Calculate the trophic efficiency if 15.8 kg of feed are needed to produce 1.0 kg of farmed bluefin tuna. Show your work. [1]

c Bluefin tuna are most commonly produced by an extensive aquaculture system. Describe *two* characteristics of extensive fish farms. [2]

d The fish waste at farms includes the parts of the tuna that are not consumed by humans, such as fins and intestines. Suggest a useful product that could be made from this waste. [1]

e Explain why supermarkets might be keen for farmed tuna to be labelled as 'dolphin friendly'. [2]

f Explain why consumers might be prepared to buy farmed tuna at a higher price. [2]

g Suggest why some consumers might consider farmed tuna to be eco-unfriendly. [2]

[Total: 11]

Chapter 9
Human impacts on marine ecosystems

CHAPTER OUTLINE

The questions in this chapter cover the following topics:
- how to use the chi-squared (χ^2) test
- how to calculate standard error and 95% confidence limits
- how to measure the gradients of lines on graph
- the effect of desalination plants on the marine environment
- the effect of acidity on coral
- the formation and negative effects of microplastics.
- bioaccumulation
- threats posed by invasive species
- coral bleaching and other consequences of the greenhouse effect
- the effects of agricultural fertiliser on oxygen concentration of water
- the importance of marine conservation.

Exercises

Exercise 9.1 Using the chi-squared (χ^2) test to compare the numbers of organisms in areas with and without desalination plants.

You need to be able to apply and use some simple statistical tests at A level. In Chapter 4 you will have learnt how to use Spearman's rank correlation to look for significant correlations. In this exercise, you will improve your understanding of the chi-squared (χ^2) test.

It is important to be able to compare numbers of individuals and frequencies. When we look at two areas, it can be obvious if there is a difference in populations of a species, but we often want to know how significant the difference is. To get an idea of how different the numbers are, we need to do a chi-squared (χ^2) test. In this exercise, you will use the chi-squared test to determine if there are significant differences between populations of mussels in areas with and without desalination plants.

> **TIP**
>
> Make sure that you know what each of the three statistical tests included in this course is used for. Spearman's rank for correlations, chi-squared for differences between frequencies or populations, and standard error with 95% confidence limits for differences between means.

9 Human Impacts on marine ecosystems

1 A student counted the number of a species of mussel present in ten one square metre quadrats in an area with a **desalination plant** and an area without a desalination plant. The results are shown in Table 9.1.

Area	Number of mussels in ten quadrats of one square metre
no desalination plant	255
desalination plant	175

Table 9.1: Populations of a mussel species in areas with and without a desalination plant.

KEY WORDS

desalination plants: industrial facilities that remove salts from seawater to manufacture fresh water

chi-squared (χ^2) test: a statistical test to measure how expected outcomes compare to actual outcomes

null hypothesis (H_0): there is no correlation between the two sets of variables

a There are different numbers of mussels in the two areas. To decide how likely it is that the difference is due to chance, we can carry out a **chi-squared (χ^2) test**. Follow the steps to determine if the difference in number of mussels in the two areas is due to chance.

Step 1: Formulate a **null hypothesis**. When we carry out a chi-squared (χ^2) test, we first start with a null hypothesis. This means that we assume that there is no association between variables. In this example, this means that there is no association between the presence of a desalination plant and the number of mussels. This should be on the lines of: 'There is no difference in the number of ....' Write down a null hypothesis for the student's experiment.

Step 2: Draw a chi-squared (χ^2) table. The observed (O) values are the results obtained from the experiment. Copy Table 9.2 and fill in the observed (O) values and the total number of mussels in both areas.

Categories	Observed (O)	Expected (E)	(O – E)	(O – E)²	$\frac{(O-E)^2}{E}$
no desalination plant	255				
desalination plant					
Total	N =	N =			$\Sigma \frac{(O-E)^2}{E} =$

Table 9.2: Chi-squared table.

Step 3: Calculate the expected (E) values. The expected (E) values are the values we would expect if there is no difference in the two areas. For our example, the total number of mussels for the two areas is $255 + 175 = 430$. If these 430 mussels were shared out equally, there would be $430 \div 2$ in each area. Calculate the expected value for each area and write them in your table.

Step 4: Complete the chi-squared table. For each row in the table, now calculate $(O - E)$, $(O - E)^2$, and $\frac{(O-E)^2}{E}$. Write the values in your table.

Step 5: Calculate the value of chi-squared (χ^2). The formula for chi-squared (χ^2) is:

$$\chi^2 = \Sigma \frac{(O-E)^2}{E}$$

This formula means 'the sum of all the $\frac{(O-E)^2}{E}$ terms.' Add up all the $\frac{(O-E)^2}{E}$ terms and add them to your table.

Step 6: Calculate the degrees of freedom. For simple chi-squared (χ^2) tests, the degrees of freedom are simply the number of categories minus 1. For our example, there are two categories (no desalination plant and with a desalination plant), which means that the degrees of freedom is $2-1=1$.

Step 7: Use a chi-squared table to determine if there is a significant difference in the mussel populations.

b   Table 9.3 is a chi-squared (χ^2) table. To use it, find the row that corresponds to the number of degrees of freedom, which is 1 in this case. The probability that we tend to look for is the column for 0.05 (5%.) A calculated value of χ^2 greater than the value for $P<0.05$ (5%) tells us that there is a probability of less than 0.05, or 5%, that the difference in the populations is due to chance. This is a statistically significant difference in the population of mussels in the two areas and we can reject our null hypothesis. If our calculated value of χ^2 is less than the critical value at $P<0.05$, there is a probability of greater than 0.05 (5%) that the difference is due to chance; in this case, the difference is not significant and we do not reject our null hypothesis.

Although we generally use $P<0.05$ (5%) as the value to decide if there is a significant difference, we can use Table 9.3 to see how significant the difference is. Columns to the right of $P<0.05$ show more significant differences and columns to the left show less significant differences. For example, if our value of χ^2 is 7.2, this is greater than the critical value for 0.01 (1%), so there is a probability of less than 0.01 (1%) that the difference is due to chance.

← decreasing significance   increasing significance →

Degrees of freedom	Probability ($P < x$)				
	0.90 (90%)	0.10 (10%)	0.05 (5%)	0.025 (2.5%)	0.01 (1%)
1	0.02	2.71	3.8	5.02	6.63
2	0.21	4.61	5.99	7.38	9.21
3	0.58	6.25	7.81	9.35	11.34
4	1.06	7.78	9.49	11.14	13.28
5	1.61	9.24	11.07	12.83	15.09

Table 9.3: Chi-squared (χ^2) critical values table.

i   Use your calculated value of χ^2 to determine if there is a significant difference in the population of mussels at the two sites. Copy out the following sentences by selecting the correct words.

'The calculated value of chi-squared (χ^2) is {*greater than / less than*} the critical value for $P<0.05$. This means that there is {*a significant difference / no significant difference*} in mussel population at the two sites. There is a {*less than / greater than*} 0.05 (5%) probability that the difference is due to chance. The null hypothesis is {*rejected / not rejected*}.'

ii Use your knowledge of the possible negative impacts of desalination plants to suggest reasons for differences in mussel population.

iii Explain how the student should have selected the area with no desalination plant and placed the ten quadrats.

2 A student investigated the impact of a tourist education campaign about the risks of plastic to the marine environment. The campaign aimed to encourage visitors to minimise their use of plastic water bottles and place the empty bottles in recycling bins. To assess the impact of the campaign, the student collected plastic water bottles washed up on a tourist beach one day before and one day after the campaign. The results are shown in Table 9.4.

Time plastic bottles collected	Number of plastic bottles collected on beach
before campaign	225
after campaign	155

Table 9.4: Data collected before and after educating tourists on the risks of plastic bottles.

a Use a chi-squared (χ^2) test to determine if there is a statistically significant difference in the number of plastic bottles collected before and after the education campaign.

b Suggest and explain how the student should have carried out the experiment to ensure validity.

c Plastic bottles can break up into **microplastics** when in the ocean. List *four* environmental factors that would increase the rate of microplastic formation.

3 Figure 9.1 shows the location of a **sewage** outflow pipe in a coastal area. The populations of oysters were measured at the East, West and South areas of the bay. The estimated populations of oyster are shown in Table 9.5.

KEY WORDS

microplastics: plastic particles that are less than 5 mm in diameter

sewage: liquid and solid waste material such as waste water or urine

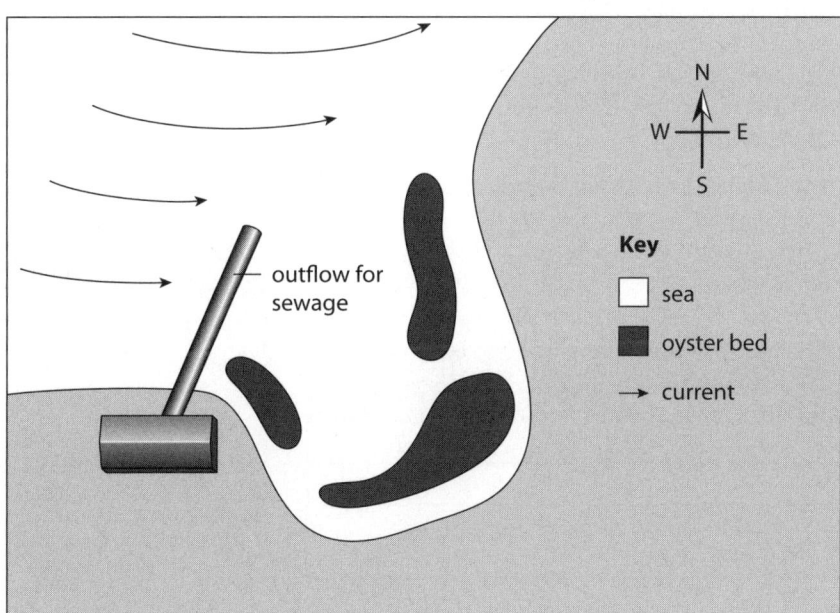

Figure 9.1: Map showing area of sewage outflow and oyster beds.

Area of oyster bed	Oyster population
East	925
West	275
South	825

Table 9.5: Oyster populations in areas of a bay near sewage outflow pipe.

a   Use the chi-squared (χ^2) test to determine if there is a statistically significant difference in oyster population in the areas of the bay.

b   Oysters are filter feeding organisms. Suggest explanations for the distributions of the oysters.

c   The oysters are fished commercially. Suggest why the oysters are monitored by fisheries agencies before they can be sold.

d   Explain how untreated sewage can lead to reduced oxygen content in water.

Exercise 9.2 Using standard error and 95% confidence limits to determine the effect of acidity and temperature on coral productivity

In Chapter 7 you used standard deviation as a descriptive measure of the variation of data about a mean. Standard deviation is useful for describing data but is not a true statistical test for determining differences between mean values. The **standard error** of a mean is a measure of how accurate a sample mean is compared to the true mean. The standard error can be used to calculate the **95% confidence limits,** which are the limits within which we are 95% sure that the true mean lies.

We can use 95% confidence limits as **error bars** on graphs – if error bars do not overlap, we can state that there is a probability of less than 5% that the difference in the means is due to chance (or alternatively, that there is a 95% probability that the difference in the means is not due to chance). This exercise will help you to understand how to calculate the standard error of a mean and 95% confidence limits by looking at the effects of **pH** and temperature on coral productivity.

1   The effects of temperature and pH due to increased carbon dioxide concentration on the productivity and growth of a species of coral were investigated. Five samples of living coral were grown at two different temperatures and three different pHs. The net productivity was measured as net oxygen production per day. The results are shown in Table 9.6.

pH	Net primary productivity at 25 °C / μmole cm^{-2} day^{-1}	Net primary productivity at 29 °C / μmole cm^{-2} day^{-1}
8.2	25, 22, 21, 20, 19, 24, 23, 18, 22, 21	9, 12, 13, 8, 8, 9, 12, 8, 9, 12
7.9	12, 13, 9, 8, 11, 13, 8, 10, 11, 9	−3, −1, −4, −1, −2, −1, −3, 0, −2, −3
7.6	1, 2, 1, 3, 0, 0, 4, 1, 1, 2	−9, −5, −7, −5, −8, −8, −9, −6, −5, −11

Table 9.6: Raw data for net primary productivity of coral at different temperatures and pHs.

> **KEY WORDS**
>
> **standard error:** a measure of how much a sample mean deviates from the true mean
>
> **95% confidence limits:** range of values between which there is 95% confidence that the mean lies
>
> **error bars:** lines through points on graphs to represent the uncertainty of a measurement
>
> **pH:** a numeric value expressing the acidity or alkalinity of a solution on a logarithmic scale

> **TIP**
>
> You can use your calculator to calculate standard deviation but make sure that you know how to do it.

a   Calculate the mean net primary productivity for the coral at each pH and temperature. Copy and complete Table 9.7.

pH	Net primary productivity at 25 °C / µmole cm^{-2} day^{-1}				Net primary productivity at 28 °C / µmole cm^{-2} day^{-1}			
	mean	standard deviation	standard error	95% confidence limits	mean	standard deviation	standard error	95% confidence limits
8.2	21.5	2.2	0.7	20.1 – 22.9				
7.9								
7.6								

Table 9.7: Mean values and standard deviations of results.

b   Calculate the standard deviations of the net primary productivities of the coral at each pH and temperature. You can either use a calculator or spreadsheet to do this automatically or use the method outlined in Chapter 7. Write your answers in your table.

c   Follow the steps to calculate the standard error and 95% confidence limits of the means.

Step 1: Use the formula to calculate the standard error:

$$\text{standard error} = \frac{\text{standard deviation}}{\sqrt{n}}, \text{ where } n \text{ is the sample size}$$

For example, for the net primary productivity at a pH of 8.2 and temperature of 25 °C, the standard error is:

$$\frac{2.2}{\sqrt{10}} = 0.7$$

Step 2: Now, calculate the standard error for all the mean values in your table. Write your answers in your table.

Step 3: Calculate the 95% confidence limits. The 95% confidence limits are the limits within which we are 95% certain that the true mean lies.

$$95\% \text{ confidence limits} = \text{mean} \pm 2 \times \text{standard error}$$

For example, for the net primary productivity at a pH of 8.2 and temperature of 25 °C, the confidence limits are:

$$21.5 \pm (2 \times 0.7) = (21.5 - 1.4) \text{ to } (21.5 + 1.4)$$

This means that we are 95% confident that the true mean lies between 20.1 and 22.9. Calculate the confidence limits for all the mean values in your table. Write your answers in your table.

d   It is usually easiest to see if there is an overlap in confidence limits by plotting a graph. Draw a bar chart to display the results of this investigation with mean net primary productivity on the vertical axis and pH on the horizontal axis. Draw bars to show each of the mean results and make sure that you add a key to clearly show the two different temperatures. Add 95% confidence limit error bars for each of the means. For example, error bars for the mean net primary productivity at pH 8.2 and temperature 25 °C will extend between 20.1 and 22.9 mmole cm^{-2} day^{-1}, as shown in Figure 9.2.

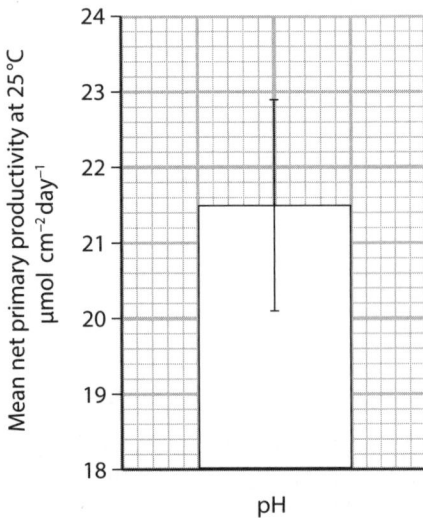

Figure 9.2: The 95% confidence limits used as error bars.

e   Interpret the results.

   Step 1: Describe the effect of pH on the mean net primary productivity of the coral by comparing the mean values for each of the three pHs at 25 °C and also at 28 °C. Describe the effect of temperature on the mean net primary productivity at 25 °C by comparing the mean values for the two temperatures at each pH.

   Step 2: Comment on whether there are statistically significant differences between means. If error bars do not overlap, there is a probability of less than 0.5 (5%) that the difference is due to chance; in other words, there is a statistically significant difference. If error bars do overlap, there is a probability of greater than 0.5 (5%) that the difference is due to chance; in other words, there may not be a statistically significantly difference in the means.

f   Explain why net production of oxygen gas was used as a measure of productivity.

g   Use your knowledge of coral bleaching and zooxanthellae to explain the results.

9 Human Impacts on marine ecosystems

2 The scientists also compared the growth of the coral by comparing the increase of mass of the coral due to calcification. They measured the percentage change in mass per month for five corals in three different pHs and two different temperatures. The results are shown in Table 9.8.

pH	Percentage change in mass per month (%) at a temperature of 25 °C	Percentage change in mass per month (%) at a temperature of 28 °C
8.2	2, 3, 2, 2, 4, 3, 3, 3, 1, 2	4, 3, 4, 2, 3, 3, 4, 2, 3, 4,
7.9	1, 4, 3, 3, 4, 2, 1, 1, 2, 3	2, 1, 1, 2, 1, 1, 2, 3, 1, 2
7.6	−1, 0, 0, 0, −2, −1, −2, 0, −1, −1	−5, −4, −2, −4, −5, −4, −4, −3, −2, −3

Table 9.8: Raw results for investigation into effects of pH and temperature on growth of coral.

a Draw a table similar to Table 9.7. For each pH and temperature, calculate the mean percentage change in mass per month and standard deviation. Write your answers in your table.

b Calculate the standard error and 95% confidence limits for the results in Table 9.8 and write them in your table.

c Draw a bar chart to show the effects of pH and temperature on the mean percentage changes in mass per month of the corals. Add 95% confidence limits as error bars.

d Discuss the results of the experiment shown by your graph. Describe the effects of pH and temperature on the growth of the corals and identify statistically significant differences in means.

e Use your knowledge of ocean acidification to explain the results of the experiment.

f Explain why increased fossil fuel combustion could cause a threat to corals.

3 The impact of a marine protected area (MPA) on the population of sharks was investigated. The MPA was established in 2005. Scientists calculated the mean number of sharks and turtles that passed a camera in 2005, 2010 add 2015. The results are shown in Table 9.10.

Date	Number of sharks of all species / sharks hr^{-1}		Number of different shark species	Mean number of turtles of all species / turtles hr^{-1}		Number of different turtle species
	mean	standard deviation		mean	standard deviation	
2005	0.15	0.08	4	0.14	0.6	3
2010	0.45	0.12	6	0.79	0.16	7
2015	0.76	0.16	11	0.54	0.20	5

Table 9.10: The effects of an MPA on numbers of shark and turtle.

a   Calculate the standard error and 95% confidence limits for the mean number of sharks of all species and the mean number of turtles of all species. In each case, the number of samples taken each year, *n*, was 12.

b   Draw a graph to show how the mean numbers of sharks and turtles change between 2005 and 2015. Add 95% confidence limit error bars.

c   Discuss the success of the MPA. In your answer, consider the significance of any changes in population.

Exercise 9.3 Calculating the rate of breakdown of plastic into microplastics by determining gradients of lines on graphs

This exercise will help you understand how to calculate the gradients of lines on graphs when analysing data. It will also improve your understanding of rates of production of microplastic and their effects on food webs.

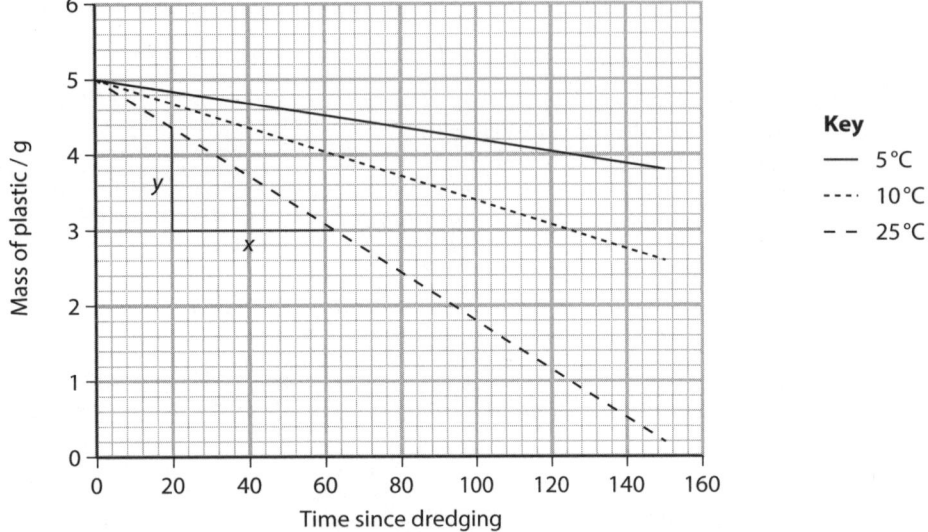

Figure 9.3: Change in mass of a 5 g plastic sheet when exposed to different water temperatures.

1   Several factors affect the breakdown of plastics into microplastics. Figure 9.3 shows the results of an experiment into the effect of temperature on the breakdown of a 5 g piece of plastic sheeting. Plastic was placed into seawater at three different temperatures and its mass measured every day.

Figure 9.3 shows the mass of plastic remaining over time after being exposed to water of different temperatures.

The gradient of a line is calculated by dividing the change in *y*-coordinate by the change in *x*-coordinate. In Figure 9.3, the gradient for the line showing the mass of plastic remaining over time with a temperature of 25 °C is:

$$\frac{y}{x} = \frac{3.7 - 2.4}{80 - 40} = \frac{1.3}{40} = 0.0325 \text{ g day}^{-1}$$

a   Calculate the gradients of the lines showing the changes in mass of the plastic at 10 °C and 25 °C.

b   Describe the effect of increasing temperature on the rate of breakdown of plastic.

c   Because you know the rates of breakdown of the plastic at each temperature, you can also give an idea of magnitude. The increase in rate as the temperature increases from 5 °C to 10 °C is:

$$\frac{\text{rate at 10 °C}}{\text{rate at 5 °C}}$$

Calculate the magnitude of increase in rate of breakdown of plastic between 5 °C and 10 °C, and between 10 °C and 25 °C.

d   Use Figure 9.3 to give the decrease in mass of a 5 g piece of plastic after 100 days when placed at a temperature of 10 °C.

e   Use Figure 9.3 to give the time taken for a 5 g piece of plastic to lose 2 g in mass when placed at a temperature of 25 °C.

2   The absorption of copper ions from surrounding seawater by two different types of plastic, labelled A and B, was investigated. Microplastic of both types of plastic was placed into a solution of seawater with copper ions. The mass of copper ions absorbed by the plastic was recorded over a 14-day period. The results for plastic A are shown in Figure 9.4.

TIP

When lines are straight, they have constant gradients so you can measure the gradients at any point. Pick sections that make it easy for you to calculate the changes in values.

TIP

If we want the know the fastest rate, we measure the gradient of the steepest tangent, which in this example is the tangent labelled p.

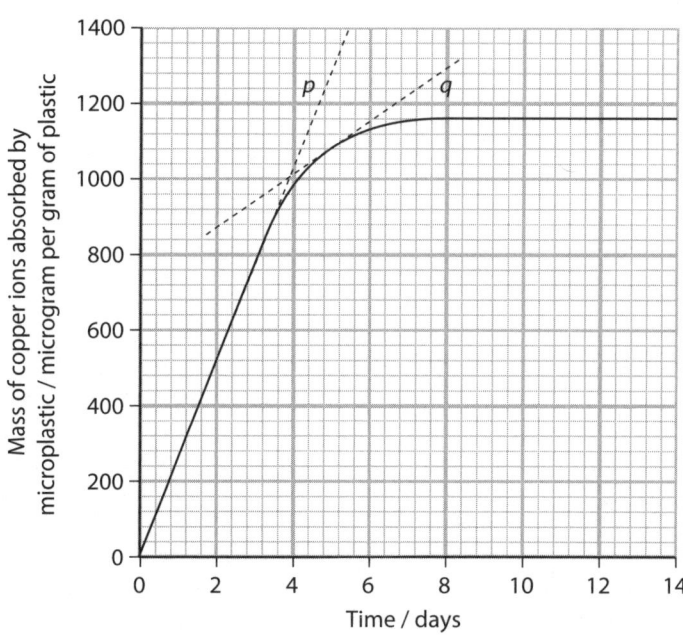

Figure 9.4: Absorption of copper ions by plastic A.

This graph is a curve of best fit. To find the rate of absorption of copper, we need to draw tangents to the curve at the point where we wish to find the rate. Figure 9.4 shows two tangents to the curve, p and q.

a   Determine the rates of absorption of copper ions by measuring the gradient of the tangents labelled p and q.

The results for plastic B are shown in Table 9.11.

Time / days	Mass of copper ions absorbed by microplastic / microgram per gram of microplastic
0	0
2	0
4	0
6	50
8	100
10	200
12	1500
14	3000

Table 9.11: Mass of copper ions absorbed by plastic B over time.

b Plot a graph to show the absorption of copper ions over time. Draw a curve of best fit through your points.

c Determine the maximum rate of absorption of copper ions by plastic B by measuring the gradient of a tangent to curve at the steepest point.

d Compare the rates of absorption of copper ions by plastic A with plastic B.

e Microplastics can be consumed by filter feeding organisms and passed through food chains. The molecules that the plastics contain are often dipolar (having charged and uncharged areas). Explain how this information and the results from this investigation suggest that microparticles may be a risk to humans and animals in higher tropic levels.

EXAM-STYLE QUESTIONS

1. **a** Figure 9.5 shows the effect of dredging on the maximum distance of light penetration and oxygen concentration of the water in an estuary over a period of time.

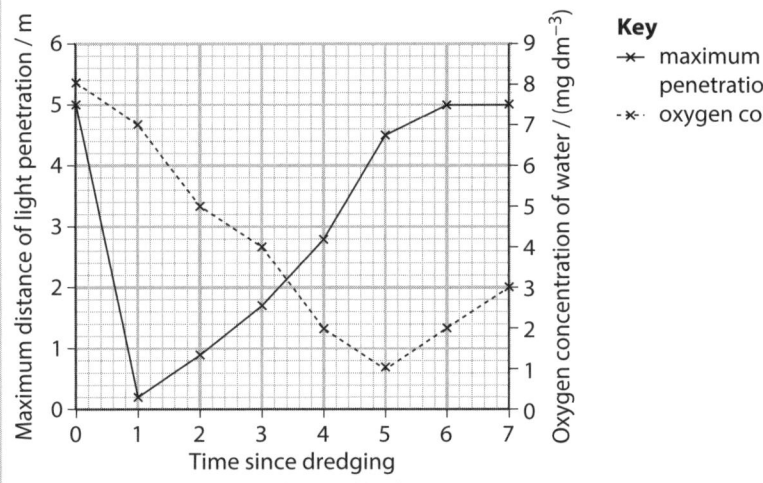

Figure 9.5: Graph to show the effect of dredging on maximum distance of light penetration and oxygen concentration of water

   i Describe the effect of dredging on the change in light penetration. [2]

   ii Calculate the percentage change in oxygen concentration of the water between day 0 and day 5. [2]

   iii Use Figure 9.5 to suggest explanations for the changes in oxygen concentration of the water. [4]

b Figure 9.6 shows a food chain found in two estuaries. One estuary is regularly dredged but the other is not. Table 9.12 shows the concentrations of mercury in the bodies of the organisms of the food chains of both estuaries.

phytoplankton → zooplankton → shrimp → trout

Figure 9.6: Diagram showing a food chain found in two estuaries

Organism	Concentration of mercury in body of organisms / µg g^{-1}	
	dredged estuary	undredged estuary
phytoplankton	0.003	0.001
zooplankton	0.05	0.02
shrimp	0.7	0.2
trout	1.3	0.8

Table 9.12: Concentrations of mercury in body of organisms found in estuaries.

CONTINUED

   i   Explain the differences in the concentration of mercury in the bodies of the organisms in the dredged estuary. [3]

   ii Suggest an explanation for the effect of dredging on the concentration of mercury in the body of organisms. [2]

   iii Suggest why pregnant women are advised to restrict the consumption of quaternary consumer fish such as swordfish. [2]

   [Total: 15]

2 Lionfish are a predatory, invasive species in areas of the Caribbean Sea. Figure 9.7 shows the change in population of lionfish in an area of the Caribbean Sea.

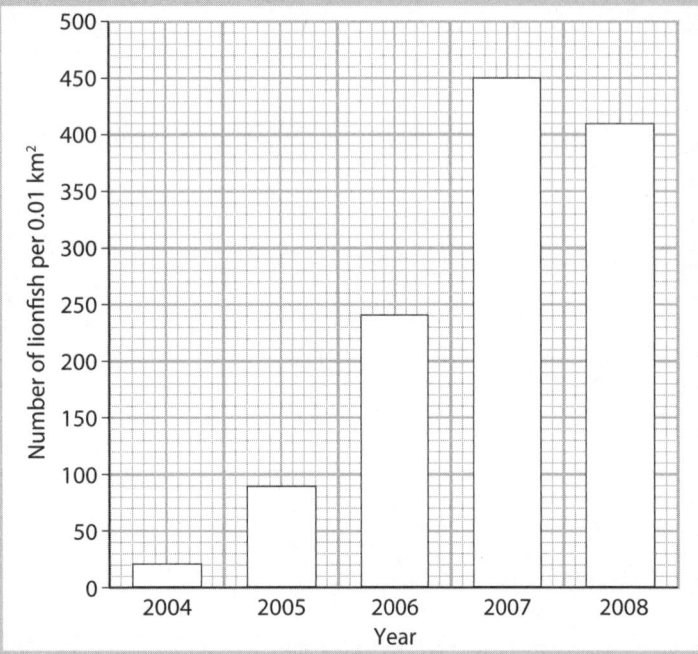

Figure 9.7: Graph to show change in population of lionfish in an area of the Caribbean Sea

   a i   Explain what is meant by the term *invasive species*. [2]

      ii Describe the changes in lionfish population from 2004 to 2008. [2]

      iii Explain the change in population of lionfish from 2004 to 2008. [3]

   b The tomtate fish is a species that is found naturally in areas of the Caribbean Sea. Young tomtate fish are eaten by lionfish, but adults are too big to be consumed. Scientists investigated the abundance of tomtate fish in an area of coral reef that had been invaded in 1997 by lionfish. Fish traps were placed around areas of a coral reef that had been invaded by lionfish. The traps were checked and the mean annual number of tomtate

CONTINUED

fish per trap was determined over time. The experiment was repeated in an area of coral reef with no lionfish. The results are shown in Figure 9.8.

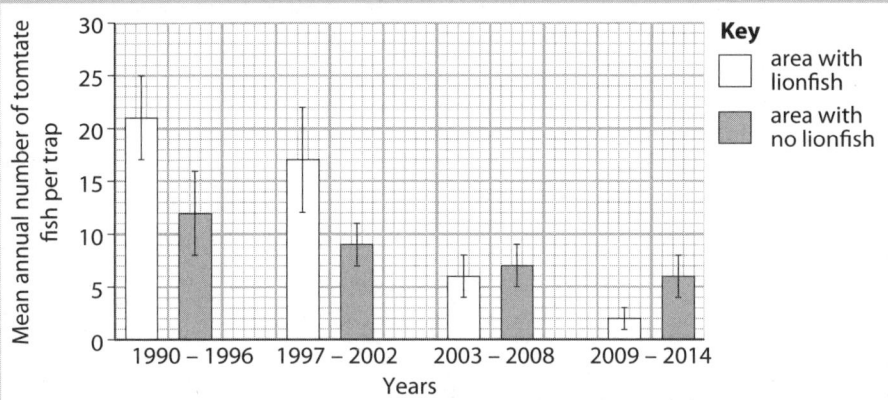

Figure 9.8: Graph to show the change in population of tomtate fish in areas with lionfish and without lionfish

    i   The error bars on the graph represent 95% confidence limits. Describe the changes in population of tomtate fish from 1990 to 2014 in the areas with lionfish. [2]

    ii   Discuss the effect of invasion of the coral reef by lionfish on the population of tomtate fish. [4]

[Total: 13]

3   Table 9.13 shows the number of coral bleaching events and the mean surface water temperature of an area of barrier reef from 1984 to 2004.

Year	Number of coral bleaching events	Mean surface water temperature / °C
1984	0	26.0
1986	0	26.5
1988	15	27.0
1990	0	26.0
1992	10	26.2
1994	10	26.3
1996	8	26.5
1998	380	27.4
2000	50	26.4
2002	360	27.4
2004	140	27.2

Table 9.13: Table showing the number of coral bleaching events and mean surface temperature of water between 1984 and 2004

CONTINUED

a i Plot a graph to show the change in number of coral bleaching events and change in mean surface temperature from 1984 to 2004. [5]

ii Calculate the percentage increase in mean surface temperature from 1984 to 2004. [2]

iii A student made the following hypothesis. 'Increasing carbon dioxide release is causing global warming that is resulting in coral bleaching.' **Evaluate** this conclusion. [4]

b Other than coral bleaching, describe possible consequences of an enhanced greenhouse effect on the marine environment. [4]

[Total: 15]

COMMAND WORD

evaluate: judge or calculate the quality, importance, amount, or value of something

4 A student investigated the effect of adding fertiliser to water on the oxygen concentration in water. They added 10 g of fertiliser to a tank of seawater containing marine algae. They monitored the mean daily concentration of oxygen in the water over a period of three weeks. The results are shown in Figure 9.9.

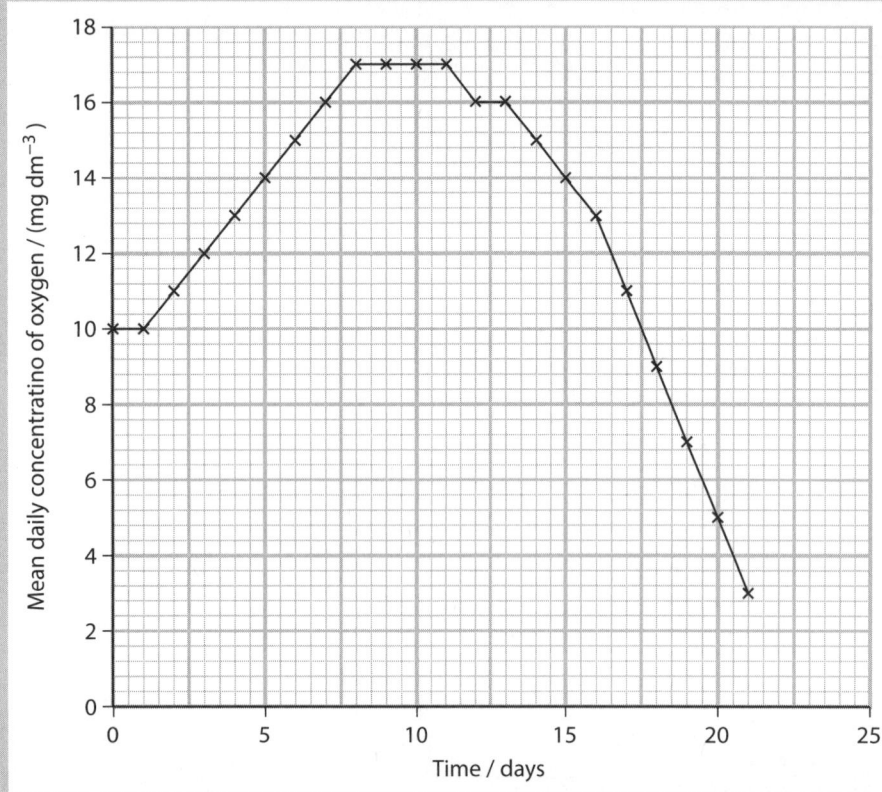

Figure 9.9: Graph to show the effect of adding fertiliser to water on the concentration of oxygen in the water

CONTINUED

    a Describe the change in oxygen of the water. [2]

    b Suggest an explanation for the effect of the fertiliser on the change in mean daily concentration of oxygen over time. [4]

    c Plan a laboratory investigation into the effect of changing the concentration agricultural fertiliser on the oxygen concentration of water. You are provided with a 10% stock of fertilizer, oxygen meter, marine algae, standard laboratory glassware and equipment. [9]

                                                                                                        [Total: 15]

5   **a** Explain the purposes of the International Union of Nature (IUCN) Red List. [4]

    b Evaluate the success of non-universal legislation such as the International Whaling Commission (IWC) moratorium and Convention on International Trade in Endangered Species (CITES) in marine conservation. [4]

    c Discuss the role of ecotourism in marine conservation. [7]

                                                                                                        [Total: 15]

AS & A Level
Section 2

SECTION OUTLINE

Marine Science is a practical subject. This section gives you an opportunity to develop a range of practical skills. It supports your understanding of many of the topics covered in the syllabus and helps you to develop the practical skills themselves, improving your ability to develop and investigate hypotheses. By carrying out all the activities in this section, you will improve your ability to plan and carry out laboratory and fieldwork activities. The activities are designed to develop and improve skills required to successfully complete investigations in Marine Science. Some of the activities in this section have been designed to help you develop surveying skills before visiting a shoreline, or to practise these skills when a visit is not possible.

Chapter 1
Water

CHAPTER OUTLINE

In this chapter you will complete investigations on:
- 1.1 properties of water
- 1.2 pH
- 1.3 salinity and temperature gradients

Practical 1.1: Properties of water

Introduction

The interaction of water molecules with each other through hydrogen bonding affects the properties of water compared with non-hydrogen-bonded liquids, such as vegetable oil. This practical explores some of these properties and demonstrates how hydrogen bonds affect these properties of water. The varying ability of water to dissolve different types of substances has important consequences for the availability of nutrients and dissolved gases to organisms that live in marine environments.

EQUIPMENT

You will need:
- vegetable oil (up to 20 cm^3 per group)
- sodium chloride (up to 10 g per group)
- glucose (up to 20 g per group)
- 2 × 10 cm^3 measuring cylinders
- 2 × dropping pipettes
- 4 × test tubes
- 1 × 50 cm^3 beaker
- test tube rack
- electronic balance
- 2 × spatulas.

Safety considerations

Take care when using glassware that test tubes or other round glass items do not roll off the worktop or break. Report any breakages to your teacher.

1 Water

BEFORE YOU START

a   We can 'measure' substances in different ways. How can a liquid be measured?

b   Liquids tend to 'stick' to the side of containers, resulting in a **meniscus** (Figure 1.1). How should you ensure that a volume of liquid is measured correctly?

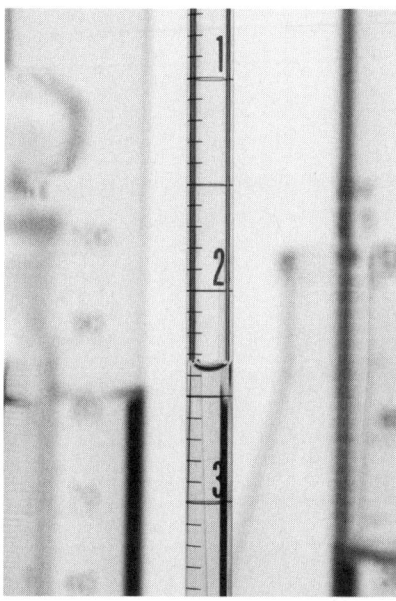

Figure 1.1: A meniscus at the top surface of water in a glass tube.

c   When using an electronic balance ensure that you use the 'zero' or 'tare' button. Why is this important?

KEY WORD

meniscus: the upward or downward curve at the surface of a liquid where it meets a container

TIP

Record quantities to the maximum precision. If the volume is $4.0\,cm^3$, record to this precision. Remember that a result of $4\,cm^3$ could have been rounded from anything as low as $3.5\,cm^3$ to as high as $4.4\,cm^3$.

TIP

To calculate density, divide the mass of liquid used by the volume the liquid occupies.

Divide mass by 1000 to convert g to kg and divide volume by 1 000 000 to convert cm^3 to $metre^3$. The density in $g\,cm^{-3}$ can be multiplied by 1000 to convert this to $kg\,metre^{-3}$.

Part 1: Density

Method

1   Place a clean, dry $10\,cm^3$ measuring cylinder onto an electronic balance, zero (tare) the balance and add 5.0 g of water to the measuring cylinder, using a dropping pipette as you get close to 5.0 g. Copy Table 1.1 and use it to record the volume of water added.

2   Place another clean, dry $10\,cm^3$ measuring cylinder onto an electronic balance, zero (tare) the balance and add 5.0 g of vegetable oil to the measuring cylinder, using a dropping pipette as you get close to 5.0 g. Record the volume of vegetable oil added.

3   Calculate the density of each liquid in grams cm^{-3}.

4   Calculate the density of each liquid in kg $metre^{-3}$.

5   Pour both the vegetable oil and water into the beaker. Does the less dense liquid float or sink?

Results

	Water	Vegetable oil
volume of 5.0 g of liquid / cm³		
density / g cm⁻³		
density / kg m⁻³		

Table 1.1: Results for density of liquids practical.

Part 2: Solvent action

Method

1. Measure 10.0 cm³ of water into each of two test tubes, labelled A and B.
2. Measure 10.0 cm³ of vegetable oil into each of another two test tubes labelled C and D.
3. Weigh out and record (in a copy of Table 1.2) the exact mass of approximately 5 g of sodium chloride.
4. Add a small amount of the sodium chloride to test tube A, shake vigorously and check to see if all the solid has dissolved. If it has all dissolved, repeat adding small additional amounts of solid and shaking until no more will dissolve.
5. Record the mass of remaining sodium chloride and calculate the amount that has been dissolved.
6. Repeat steps 3–5 with:
   - 10 g of glucose in test tube B (water).
   - 5 g of sodium chloride in test tube C (vegetable oil).
   - 10 g of glucose in test tube D (vegetable oil).

Results

Test tube	Mass of solid at start / g	Mass of solid remaining / g	Mass of solid dissolved / g
A			
B			
C			
D			

Table 1.2: Results for Solvent action practical.

Evaluation and conclusions

d   How does the density of water compare to the density of vegetable oil? How does this explain what happens when water and vegetable oil are mixed together?

e   Is density the same for a certain substance at different temperatures? How could you investigate this?

f Salt is an ionic substance; glucose is covalent. Explain the solubility results in terms of the interactions between solvent molecules and solute particles.

g Outline an investigation to test this hypothesis: 'The solubility of a salt depends on the temperature of the water.'

h How does the ability of water in the oceans to absorb heat energy affect the rate at which global warming heats Earth's atmosphere?

Reflection

i How could the results for the solubility experiment be improved?

j Compare your results for both parts of the practical to the results of others in the class. How do your results compare? Suggest reasons for any differences among your results and how these could be minimised if you repeated the practical.

Practical 1.2: pH

Introduction

Whether a substance is neutral, acid or alkaline depends on the concentration of hydrogen ions, H^+. The more hydrogen ions there are in solution, the more acidic the solution is.

The pH scale is a logarithmic scale. This means that for each decrease in pH of 1 there are 10 times more hydrogen ions present. So, small changes in pH are due to large changes in the concentration of hydrogen ions present.

The pH of seawater is important because it affects the ability of organisms to carry out essential biochemical reactions. The aim of this practical is to show how small changes in pH represent a significant change in the concentration of hydrogen ions in water.

EQUIPMENT

You will need:

- 1.0 M hydrochloric acid (up to 30 cm³ per group)
- 1.0 M sodium hydroxide solution (up to 30 cm³ per group)
- universal indicator solution or paper with corresponding colour pH chart
- litmus indicator solution (or red and blue litmus papers)
- pH meter
- 6 × test tubes or boiling tubes (need to be wide enough to allow the pH probe to be used)
- test tube rack
- distilled water (up to 20 cm³ per group for testing, plus extra for rinsing boiling tubes and pH probe)
- seawater (up to 20 cm³ per group)
- carbonated water (for example, bottled sparkling water) (up to 20 cm³ per group)
- dropping pipettes.

Safety considerations

1.0 M sodium hydroxide is corrosive and particularly dangerous to eyes. You must wear eye protection. Universal indicator solution – check risks of solution provided. Many are highly flammable and some are harmful.

> **BEFORE YOU START**
>
> a   Why is it important for the pH probe to be calibrated before you record measurements?
>
> b   Universal indicator is a mixture of several different indicators. The combination of indicators varies between different manufacturers. Why is it important that you use the correct colour chart for the actual indicator you use?
>
> c   You will need to re-use your boiling tubes during this experiment. What steps will you take to ensure that all your results are accurate?

> **TIP**
>
> When recording colour changes, it is useful to hold the solution against a white background, such as a plain white sheet of paper or a white tile.

Part 1: Different methods of measuring acidity and pH

Method

1. Pour approximately 1 cm depth of 1.0 M hydrochloric acid into a boiling tube.
2. Test the solution with a few drops of universal indicator solution or universal indicator paper. In a copy of Table 1.3, note the resulting colour of the indicator with the corresponding pH from the colour chart.
3. Repeat steps 1 and 2 with: 1.0 M sodium hydroxide solution, distilled water, seawater and carbonated water.
4. Repeat the tests on new samples of each solution using litmus indicator and then a pH probe. Record all your results in a copy of Table 1.3.

> **TIP**
>
> Record all numerical results to the same number of decimal places to show that you have recorded them as precisely as possible (for example, record pH 4.0 not pH 4).

Results

	Colour and pH of universal indicator	Colour of litmus indicator	pH probe reading
1.0 M hydrochloric acid			
1.0 M sodium hydroxide			
distilled water			
seawater			
carbonated water			

Table 1.3: Results table for measuring acidity and pH.

Part 2: Concentration of hydrogen ions and the pH scale

Method

1. Label six boiling tubes with numbers from 1 to 6.
2. Measure $10.0\,cm^3$ of $1.0\,M$ hydrochloric acid and pour this into tube 1.
3. Remove exactly $1.0\,cm^3$ of the solution from tube 1 using a dropping pipette and add this into tube 2. Add $9.0\,cm^3$ distilled water and mix gently but thoroughly.
4. Remove exactly $1.0\,cm^3$ of the solution from tube 2 using a dropping pipette and add this into tube 3. Add $9.0\,cm^3$ distilled water and mix gently but thoroughly.
5. Repeat this diluting process until you have solutions in all six boiling tubes.
6. Test the pH of each solution with the pH probe and record your results in a copy of Table 1.4.

Results

Test tube	pH
1	
2	
3	
4	
5	
6	

Table 1.4: Results for diluting a sample of acid and measuring pH.

Evaluation and conclusions

d What are the advantages and disadvantages of each method of measuring acidity?
e Why would the pH of the acid not increase above 7 with further dilution?
f What do you predict would happen to the pH of the carbonated water over time? Write a hypothesis that you could test. Outline an experiment you could carry out to test your hypothesis.

Reflection

g Compare your results with those from other groups. How similar are the results?
h How could the accuracy of the dilutions in Part 2 be improved?

Practical 1.3: Salinity and temperature gradients

Introduction

You have learnt how density is measured and how this varies for different substances. Density can also vary in the same substance depending on the temperature or amount of solutes dissolved in it. This is important in marine environments where the surface of oceans is heated by energy from the sun, and where rivers meet the sea bringing fresh water into the more saline seawater.

EQUIPMENT

You will need:

- salt water (40 g dm^{-3}) (approx. 40–50 cm^3 per group)
- distilled water (up to 50 cm^3 per group)
- 2 contrasting food colours
- 2 × 100 cm^3 beakers
- dropping pipette
- hot water (students or the teacher could produce this).

Safety considerations

Take care using hot water because both steam and boiling water can cause burns. Beakers containing hot water will also quickly get very hot.

BEFORE YOU START

a Explain the meaning of the terms *thermocline* and *halocline*.
b How can you reduce the risk of handling hot water?

Part 1: Temperature gradients

Method

1 Add a small amount of food colouring to about 50 cm^3 of cold tap water in the first beaker to give a distinct colour and mix well.

2 In a second beaker add a small amount of a different food colouring to about 50 cm^3 of hot water from a kettle to give a distinct colour and mix well.

3 Make sure that the cold water has stopped swirling in the beaker.

4 Use a dropping pipette to carefully transfer about 1–2 cm^3 hot water. Insert the tip of the pipette into the middle of the cold water and gently release the hot water into the cold tap water. Observe what happens to the hot water.

5 Repeat this process until 15–20 cm^3 of hot water have been transferred.

Results

Sketch a diagram to show any layers that have formed. Label the layers clearly to identify which is hot water and which is cold water, and (if appropriate) where some mixing has taken.

Part 2: Salinity gradients

Method

1. Add food colouring to about 50 cm^3 of salt water in the first beaker to give a distinct colour and mix well.
2. In a second beaker add a different food colouring to about 50 cm^3 of distilled water to give a distinct colour and mix well.
3. Make sure that the salt water has stopped swirling in the beaker.
4. Use a dropping pipette to transfer about 1–2 cm^3 distilled water. Insert the tip of the pipette into the middle of the salt water and gently release the distilled water into the salt water. Observe what happens to the distilled water.
5. Repeat this process until 15–20 cm^3 of distilled water have been transferred.

Results

Sketch a diagram to show any layers that have formed. Label the layers clearly to identify which is salt water and which is fresh water, and (if appropriate) where some mixing has taken place.

Evaluation and conclusions

c   How does **salinity** affect the density of water?
d   How does temperature affect the density of water?
e   What would happen if ice (frozen water) were added to cold water?
f   Outline an investigation you could carry out to produce a line graph showing how temperature affects the density of water.
g   What factors may cause different layers of water to mix?
h   Suggest how changes in the density of water, such as changes caused by rapid cooling at the surface, would affect the movement of water in the ocean.

> **KEY WORD**
>
> **salinity:** a measure of the quantity of dissolved solids in ocean water, represented by parts per thousand, ppt or ‰

Reflection

i   How has this practical helped you to understand the formation of thermoclines and haloclines?

Chapter 2
Earth processes

CHAPTER OUTLINE

In this chapter you will complete investigations on:
- 2.1 Investigating the effect of temperature on the solubility of a salt
- 2.2 Modelling weathering and erosion
- 2.3 Interpreting tide tables

Practical 2.1: Investigating the effect of temperature on the solubility of a salt

Introduction

Most minerals and salts dissolve better at higher temperatures. This is important in the formation of hydrothermal vents where hot water dissolves minerals which then precipitate back as solids when the water cools after it exits the vent. This investigation compares the solubility of a salt, by progressively diluting a concentrated solution of ammonium chloride and recording the temperature at which it begins to recrystallise as it cools.

EQUIPMENT

You will need:
- 1 × boiling tube
- boiling tube rack
- 2 × 250 cm^3 beakers
- 1 × stirring thermometer (−10 °C to 110 °C) (or a stirring rod and a thermometer)
- 1 × 10 cm^3 measuring cylinder
- dropping pipette
- boiling tube holder
- 3.0 g ammonium chloride solid
- electronic balance
- ice
- hot water.

Safety considerations

Ammonium chloride is harmful if swallowed and an irritant to the eyes. Take care stirring with the thermometer. Your teacher may advise you to use a stirring rod instead to ensure the solution heats and cools uniformly. Take care with hot water as this can burn. Wear eye protection. Some types of glass may break when they are heated or cooled quickly. Check with your teacher about clearing up broken glass.

2 Earth processes

> **BEFORE YOU START**
>
> **a** Several methods of setting up a hot water bath may be used. Describe two different methods of setting this up and compare the risks for each method.
>
> **b** Ammonium chloride is more soluble than most minerals that form hydrothermal vents. Suggest why ammonium chloride is being used for this experiment, rather than a mineral that does actually form hydrothermal vents.

Method

1. Set up a hot water bath in one of the beakers and an ice bath in the other beaker.
2. Weigh out 3.0 g of ammonium chloride and add this to the boiling tube.
3. Add 5.0 cm³ of water to the boiling tube and warm the contents in the hot water bath until all the solid dissolves.
4. Transfer the boiling tube to the ice bath and stir with the thermometer (or stirring rod).
5. When crystals start to form, note the temperature at which this occurs in a copy of Table 2.1.
6. Add another 1.0 cm³ of water to the boiling tube and warm the contents again in the hot water bath until all the solid dissolves.
7. Transfer the boiling tube to the ice bath and stir with the thermometer (or stirring rod).
8. When crystals start to form note the new temperature at which this occurs in a copy of Table 2.1.
9. Repeat steps 6–8 until a total of 12 cm³ of water have been used.

Results

Volume of water / cm³	Concentration / g dm⁻³	Temperature at which crystals appear / °C
5.0	600	
6.0	500	
7.0	429	
8.0	375	
9.0	333	
10.0	300	
11.0	273	
12.0	250	

Table 2.1: Table of results for solubility of a salt practical.

> **TIP**
>
> The concentrations have been calculated for you. They are calculated by dividing the mass of solute (3.0 g) by the volume of water (in dm³, which is the number of cm³ divided by 1000).

Evaluation and conclusions

c Plot a graph of your results, selecting the most appropriate data for your axes. Draw a suitable line of best fit for your results.

d Describe the pattern in your results. What conclusion can you draw from your results?

e Are there any **anomalies** in your results? How do you know? Suggest why these may have occurred.

f How do the results from your experiment help to explain the formation of hydrothermal vents?

g Suggest why hydrothermal vents at different locations differ in appearance and mineral composition.

Reflection

h How could you adapt the method to gain more evidence and make your results more reliable?

Practical 2.2: Modelling weathering and erosion

Introduction

Weathering and **erosion** are separate processes that are commonly confused. Weathering is the breaking down of rocks into smaller fragments (this can be physical, chemical or organic), while erosion is the removal of fragments of rock from their location (this can be by ice, water, wind or gravity). In this practical, you will model some of the processes involved, to help you to understand the differences between weathering and erosion.

EQUIPMENT

You will need:

- old glass jar or bottle, with a lid
- plastic bag large enough to seal the glass container inside
- plastic bowl
- medium (3–10 cm diameter) sample of **sedimentary rock** (such as sandstone or limestone)
- 2 × dropping pipettes
  - access to a freezer
- small (1–5 cm diameter) sample of **carbonate rock** (such as marble or limestone)
- beaker (big enough to contain the carbonate rock)
- $1.0\,mol\,dm^{-3}$ hydrochloric acid (up to $20\,cm^3$ per group)
- large rectangular plastic container (between 2 litres and 10 litres in size)
- dry sand (enough to fill approximately one-fourth of the plastic container)
- several small pebbles (1–5 cm diameter)
- straw or tube to blow through
- jug to transfer water
- large rectangular tray.

KEY WORDS

anomaly: a result or observation that deviates from what is normal or expected; in experimental results, it normally refers to one repeated result that does not fit the pattern of the others

weathering: the wearing down or breaking of rocks through physical, chemical, or organic means

erosion: a natural process where material is worn away from the Earth's surface and transported elsewhere

sedimentary rock: rock formed by the deposition of particles on the ocean floor

carbonate rock: a rock with a major component of minerals containing carbonate ions (for example, limestone contains mostly calcium carbonate)

Safety considerations

Wear eye protection throughout the investigation to avoid getting sand, rock fragments or acid in your eyes. Follow any further instructions as directed by your teacher (for example, avoid cuts by not touching any broken glass, and use disinfectant solutions before blowing through straws or tubing).

> **BEFORE YOU START**
>
> a Think about sanding the corners of a block of wood. Is the abrasion from sandpaper like erosion or like weathering?
>
> b After you sand the block of wood, what happens when you blow air at the block and sawdust around it? Which process is this more like?

Part 1: Physical weathering

Method

1. Fill an old glass jar or bottle with water and seal with a lid.
2. Place the sealed glass container in a plastic bag in a plastic bowl.
3. Take a dry sample of sedimentary rock, such as sandstone or limestone, and slowly add drops of water using a dropping pipette to the top surface of the rock until no more appears to be absorbed.
4. Place the rock in the bowl with the bag containing the glass container. Record the appearance of both the bottle and the rock in a table like Table 2.2.
5. Place the bowl in a freezer overnight.
6. The following day, remove the bowl from the freezer and observe the glass container and the rock. Record any differences in the appearance of both items in your results table.

> **TIP**
>
> Observe the overall appearance of the surface and shape of the bottle and rock. After freezing, compare the shape and surface and note any smaller fragments if there are any.

Results

Item	Appearance before freezing	Appearance after freezing
glass container		
rock containing water		

Table 2.2: Results table to record observations from Part 1.

Part 2: Chemical weathering

Method

1. Take a sample of a carbonate rock (such as marble or limestone) and place it in the beaker.
2. Use a dropping pipette to slowly add hydrochloric acid to the rock. Observe what happens to the rock.

3   Repeat until at least 10 cm^3 of acid has been added. Look at the liquid at the bottom of the beaker. Note in a table like Table 2.3 how this appears compared to the amount of acid that was added.

Results

Item	Observations and changes when acid is poured on the rock
surface of rock	
acid / liquid in beaker	

Table 2.3: Results table to record observations from Part 2.

Part 3: Wind erosion

Method

1   Place the dry sand in one end of the container to create a sloped sand 'dune' and gently push the pebbles into the surface of the slope (see Figure 2.1).

2   Make sure to wear goggles. Use a straw or flexible tube to gently blow air directed at the dune.

3   Record your observations in a table like Table 2.4. Note whether the sand and pebbles or boulders moved very far and how their distribution changed.

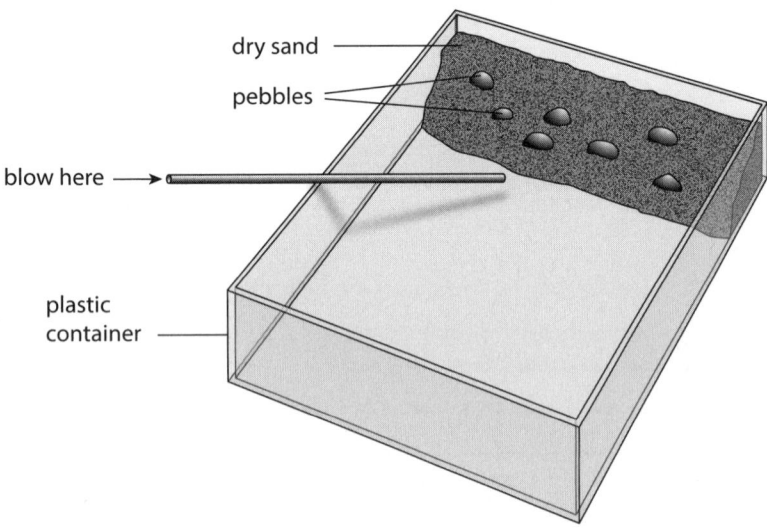

Figure 2.1: Diagram showing how to model wind erosion on a sandy shore.

Results

	Observations and changes when air is blown at the model of a 'shore'
sand	
pebbles / boulders	

Table 2.4: Results table to record observations from Part 3.

Part 4: Water erosion

Method

1. Wet the sand and create a wide shoreline on the tray. Tilt the tray so that the bottom edge hangs over the plastic container used in Part 3 (see Figure 2.2).

2. Using your finger, or the end of a stirring rod or similar instrument, create a shallow 'riverbed channel' in the shore from the top edge of the shore to the lowest edge (see Figure 2.2). This should be approximately 1 cm deep.

3. Using a jug, gently pour water into the top of the 'riverbed' so that the river fills with water and flows down to the plastic container below. Observe what happens to the riverbed and the mouth of the river as more water flows through. Try gently increasing the flow of water in the river by adding water more quickly. In a table like Table 2.5, record how both the 'river' and the 'river mouth' change as more water flows.

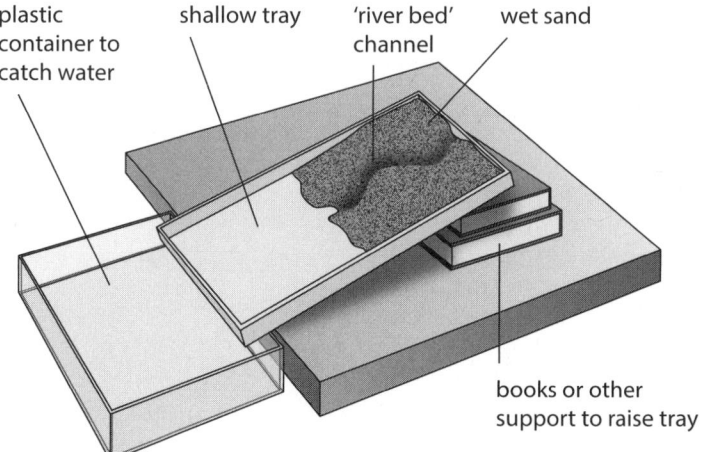

Figure 2.2: Diagram showing how to model water erosion from a river.

4. Carefully pour away any water from the plastic container (avoid letting any sand go down the drain). Create a sand dune using the wet sand from the tray and some pebbles like those you used in Part 2. Then add more water to the container to create a 'sea' (see Figure 2.3).

5. Take a plastic sheet that fits into the container and use this to gently create and push waves towards the beach. Record how the wave action changes the shoreline in a copy of Table 2.5 (compare the effects of the wave action on the sand and pebbles, and how the shape of the shore changes).

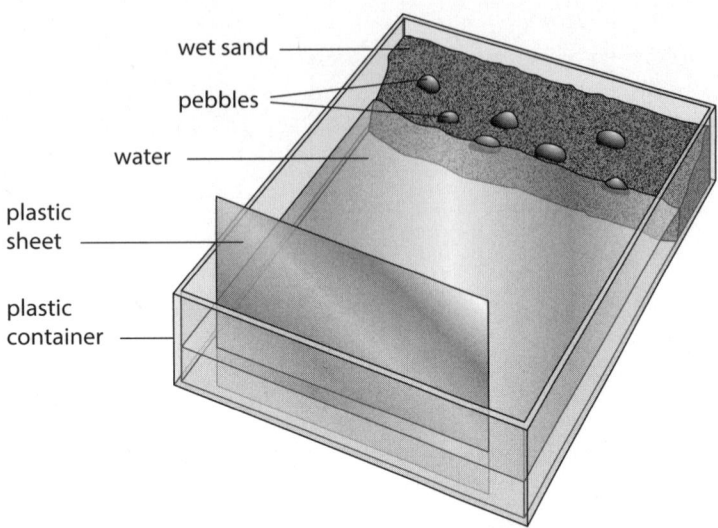

Figure 2.3: Diagram showing how to model water erosion due to wave action.

Results

	Observations and changes
riverbed with water flowing	
river mouth with water flowing	
shape of shore / beach as waves hit repeatedly	

Table 2.5: Results table to record observations from Part 4.

Evaluation and conclusions

c What happens to ice as it freezes? How does this cause the changes that you observed in Part 1 of the practical? Suggest how repeated freezing and thawing over time can lead to large rocks crumbling into much smaller sediments.

d How does the chemical weathering in Part 2 of the practical differ from the physical weathering in Part 1? What changes have occurred in the rocks in each case?

e Organic or biological weathering is a third type of weathering that can occur, such as when roots of plants grow into small cracks in rocks. Suggest how this is both similar to and different from the physical weathering in Part 1 of the practical.

f Compare the actions of both wind and water on the model beaches in Parts 3 and 4 of the practical. To what extent are these types of erosion similar and different?

g How does the size of rock particle affect the rate of erosion?

h What happens to eroded particles that are removed?

i Suggest how increased rainfall and stormy weather will affect the rate of erosion that takes place.

j Glaciers are large bodies of ice that slowly move down valleys under their own weight due to gravity. Suggest how the movement of glaciers will affect both the rate of erosion and the size of particles eroded compared to the movement of rivers.

k On a warm dry day, the edges of sandy cliffs can often be seen to crumble and fall due to gravity. Would this be described as weathering or erosion? Explain your answer.

Reflection

l How has this activity helped you to understand the difference between weathering and erosion? To what extent do you think it accurately models these changes?

m Suggest how the models could be improved.

Practical 2.3: Interpreting tide tables

Introduction

Tide tables are available for many coastal locations around the world. The height and timing of the tides are caused by the interaction of gravitational attractions between the Earth, Moon and Sun. In fieldwork at coastal locations, it is important to be able to predict the times and heights of the tide so that the fieldwork can be completed successfully and safely. It is also important to monitor the actual height of the tide during an activity (for example, using a height marker in the water, as in Figure 2.4). This activity uses tide information to produce a graph showing how the tidal range changes over a month, and relates these changes to the rotation of the Moon around the Earth and alignments with the Sun.

Figure 2.4: A height marker used to determine the actual height of the water in a harbour.

EQUIPMENT

You will need:
- tide information for the next month
- information about dates of Moon phases over the next month
- graph paper
- pencil
- eraser.

Safety considerations
There are no significant safety issues associated with this practical investigation.

> **BEFORE YOU START**
>
> a How does the tidal range relate to the phases of the Moon (that is, new moon, first quarter, full moon and last quarter)?
>
> b What other factors affect the height of the tide experienced at any location?
>
> c What is meant by the term tidal range?

> **TIP**
>
> A tide-forecast website is a good source of tidal information for many coastal locations around the world, as well as dates of Moon phases.

Method

1 Look at the tidal data for your chosen location. Identify the heights of highest high tide and the lowest low tide for the full period of time shown. Use this information to plan the scale for the y-axis (height of tide) for your graph.

2 Plot each successive high tide and low tide for the full month. Join the plots for all the high tides using a ruler and pencil. Join the plots for all the low tides using a ruler and pencil.

3 Mark on your graph the dates of: new moon, first quarter, full moon and last quarter.

> **TIP**
>
> For simplicity, the x-axis (time) will not have units, but you should mark the dates as appropriate along this axis as you plot your points.

Evaluation and conclusions

d From your graph, identify the two periods of time with the greatest tidal range (**spring tides**). Which phases of the Moon correspond to these tidal ranges? Which phase had the greatest tidal range?

e From your graph identify the two periods with the smallest tidal range (**neap tides**). Which phases of the Moon correspond to these tidal ranges?

f Explain how the relative alignment of the Sun, Earth and Moon cause a spring tide.

g Explain how the relative alignment of the Sun, Earth and Moon cause a neap tide.

h Explain why some locations such as the Bay of Fundy (between Nova Scotia and New Brunswick in Canada) or the Irish sea (between the United Kingdom and Ireland) experience much higher tides than most locations exposed to major oceans.

> **KEY WORDS**
>
> **spring tide:** a tide that occurs when the Sun and Moon are aligned, causing the largest tidal range
>
> **neap tide:** a tide that occurs when the Moon and Sun are at right angles from each other, causing the smallest tidal range

Reflection

i Discuss how this activity has helped you to understand how and why the tides change over a month.

> Chapter 3
Interactions in marine ecosystems

CHAPTER OUTLINE

In this chapter you will complete investigations on:

- 3.1 Pyramids of numbers and biomass
- 3.2 Planning an investigation to estimate the productivity of an aquatic producer
- 3.3 Investigating the carbon cycle

Practical 3.1: Pyramids of numbers and biomass

Introduction

Food chains and **food webs** show feeding relationships between species at different trophic levels. Producers are the basis of all food chains and they generally use photosynthesis to convert light energy into chemical energy and store some of this as biomass. Some of this energy and biomass is passed along food chains to higher trophic levels. This activity attempts to measure both the number and biomass of organisms at each trophic level in a habitat – ideally a shoreline habitat where this is accessible (for example, a rocky or sandy shore where you might expect to find organisms such as those shown in Figure 3.1), but a terrestrial habitat such as open grassland or woodland would also work in demonstrating how pyramids of numbers and pyramids of biomass compare.

KEY WORDS

food chain: a way to describe the feeding relationships between organisms

food web: a way to show all the different feeding relationships in an ecosystem

Figure 3.1: Examples of organisms you might find on a rocky shore.

> CAMBRIDGE INTERNATIONAL AS & A LEVEL MARINE SCIENCE: WORKBOOK SECTION 2

EQUIPMENT

You will need:

- quadrat (for example, 1 metre × 1 metre or 0.5 metre × 0.5 metre)
- plastic beakers or tubs
- trowel or small spade
- forceps
- scissors
- **pooter** (or similar device for collecting small invertebrates safely)
- large white sorting tray
- identification key
- hand lens
- balance.

KEY WORD

pooter: a bottle for collecting small invertebrates, having one tube through which they are sucked into the bottle and another, protected by muslin or gauze, which is sucked

Safety considerations

Always work with a partner for fieldwork, especially on shorelines. Depending on the habitat chosen, there may be specific hazards due to the location or nature of organisms that may be encountered. Follow all instructions from your teacher carefully.

BEFORE YOU START

a How might the location of organisms on a rocky shore differ from those on a sandy shore? How would this change your approach to collecting all the organisms present in a sample of the habitat?

b What safety precautions should you take when handling unidentified organisms?

c Why is it important to handle living organisms with care?

Method

1 Select an area in the habitat that contains a range of organisms. You may need to move seaweed or a plant cover to see what is beneath.

2 Place your quadrat carefully in the area so that no organisms are harmed.

3 Collect all the organisms in the area (cut seaweeds at the base or include the roots of plants) and place all these on the white tray. Where the substrate is loose or soft (for example, sand or soil) you could gently dig the top 10–15 cm to locate other organisms hiding in the substrate.

4 Carefully separate any animals from the producers using the forceps and place each species into a separate plastic container. Very small organisms may be easier to separate using a pooter.

5 Use the identification keys to try to identify the animals you have found. Label them according to which trophic level they occupy (primary consumers or secondary consumers).

6 Prepare a results table to record all the data you collect.

7 Count all the producers and record the number in your results table.

8 Weigh the mass of all the producers and record this in your results table.

9 Repeat steps 6 and 7 for each trophic level (not individual species) that you have found.

10 Return all the animals to their habitat.

Results

d Draw a results table to record all your results. Make sure that you have columns for both the number of organisms and the mass of the organisms.

Evaluation and conclusions

e Use graph paper to construct a pyramid of numbers and a pyramid of biomass for your results. Use the grid to draw the bars in your pyramid to scale. Label your bars with the trophic levels and include a suitable scale with units.

f If you had more time available, you could have repeated the experiment several times. What would be the advantages of this? What ecological problems might this create?

g Based on your pyramid of biomass, calculate the energy losses through these food chains. Suggest reasons for the loss of energy in the food chains.

h Suggest reasons why the data collected may not accurately reflect the biomass in each trophic level for the community sampled.

i A pyramid of number or biomass usually considers a single food chain. Suggest reasons why it would be difficult to carry out this investigation to look at a single food chain.

Reflection

j To what extent do the results from your investigation match what you expected the results to be? If they are very different, suggest why that might be.

Practical 3.2: Planning an investigation to estimate the productivity of an aquatic producer

Introduction

Photosynthesis captures light energy and transfers this into chemical energy that producers can use and store as chemical energy. The energy that is stored accumulates as biomass in the producer and becomes available to other trophic levels through feeding relationships in food chains. In this activity, you will plan and carry out an investigation to try to determine both the net productivity and the gross productivity of an aquatic plant.

Note: Gross vs Net productivity is an extension exercise only and is not part of your syllabus.

BEFORE YOU START

a When plants photosynthesise more quickly than they respire, would you expect their biomass to increase or decrease?

b Suggest how an increase or decrease in biomass of a producer could be measured.

Planning

Primary productivity can be determined by this formula:

> gross primary productivity = net primary productivity + respiration

in which:

> gross primary productivity = total energy captured by the producer
>
> net primary productivity = energy transferred into new biomass
>
> respiration = energy used by the producer

Variables

1 Copy and complete a table like Table 3.1 to identify variables that may affect the rate of growth of aquatic plants.

Factor affecting growth of plant	Why it affects growth of plant	How increasing it will affect the growth of plant	How could you change it in your practical?
light intensity			
temperature			
surface area of water			
availability of nutrients			
mass of plant at start			

Table 3.1: Factors affecting the rate of growth of an aquatic plant.

> **TIP**
>
> Energy can be estimated by measuring changes in biomass of an organism. Dry biomass readings are the most accurate, but because the plants are aquatic their water content should be consistent, so that wet biomass will give a reasonable indication of change.

2 To produce valid data, you need to change one variable (the independent variable) and measure another (dependent variable). To estimate productivity, you also need to include a measure of the time taken as well as the amount of growth.

  i   To determine the gross primary productivity what will you need to measure (dependent variable)?

  ii How will you measure your dependent variable?

  iii How many times will you repeat each measurement?

  iv What will be the independent variable (what are you changing?)

  v   How will you change your independent variable?

3 To produce valid data, you must try to control other variables that may affect your results. These are called **control variables**. Use the information you completed in Table 3.1 to identify control variables and how you can keep these constant in your investigation. Create and complete a table like Table 3.2.

> **KEY WORD**
>
> **control variables:** variables that are not being tested but that must be kept the same in case they affect the experiment

Variable	How to keep it constant

Table 3.2: Control variables for productivity investigation.

Equipment

4   List the equipment you will need to complete your investigation in a table like Table 3.3.

Equipment	Quantity

Table 3.3: Equipment needed for productivity investigation.

Safety considerations

5   What hazards might your method involve? How can you reduce the risk from these hazards? Create and complete a table like Table 3.4

Hazard	Steps taken to reduce the risk from this hazard

Table 3.4: Risk assessment for productivity investigation.

Method

6   Write a full step-by-step method of how to carry out the experiment. Include a diagram if this helps explain how to set up apparatus. Include all the practical details you need to carry out your experiment, including how to change the independent variable, how to measure the dependent variable and how to keep the control variables the same.

7   Check your plan with your teacher. Once it has been agreed, carry out your investigation.

Results

8   Draw a results table to record the data you plan to collect.

> **TIP**
>
> Make sure you include spaces for all the measurements you plan to take, including repeats and means. Ensure that all appropriate units are included in headings.

Evaluation and conclusions

You should have three measurements:

   Increase in biomass of plants exposed to light (net primary productivity) =

   Change in biomass of plants kept out of light (respiration) =

   Time taken for changes in biomass =

c   Use the formula: gross primary productivity = net primary productivity + respiration to calculate the gross primary productivity for this time period.

d   Divide your answer by the time taken to estimate the rate of gross primary productivity (make sure you include all the units).

e Suggest how the rate of productivity might vary during the year and why you cannot assume your rate of productivity is constant through the year.

Reflection

f How could you have improved your method to gain a more accurate estimation of productivity during a full year?

Practical 3.3: Investigating the carbon cycle

Introduction

The **carbon cycle** involves a number of chemical processes that convert carbon to and from different chemical forms. This series of practical activities is designed to improve your understanding of the different chemical forms of carbon and how they change into one another.

> **KEY WORD**
>
> **carbon cycle:** the range of processes that involve the chemical and physical changes to carbon resulting in carbon transforming through a range of substances, including in the atmosphere, living organisms and rocks

EQUIPMENT

You will need:

- small piece of chalk or limestone
- 1.0 mol dm^{-3} hydrochloric acid (approx. 20 cm^3)
- dropping pipette
- boiling tube fitted with a bung containing a delivery tube (see Figure 3.2)
- test tube
- limewater (approx. 20–30 cm^3)
- 2 × 100 cm^3 beakers
- universal indicator solution (up to 5 cm^3)
- straw or tube to blow through (with disinfectant solution if this is being re-used)
- seawater (approx. 50 cm^3)
- glass funnel, boiling tube and rubber bung connected by delivery tubes (see Figure 3.3)
- filtering pump or other pump suitable for suction filtration
- paraffin wax tea light candle
- small pieces of wood (for example, wooden splints)
- tongs.

Safety considerations

Wear eye protection when carrying out these experiments. Limewater (calcium hydroxide solution) can be an irritant. Take care to avoid getting this in your eyes and wash any splashes off your skin immediately. Before blowing through a straw or tubing, make sure that it is sterile or has been disinfected, and take care not to suck any solutions into your mouth.

BEFORE YOU START

a In Practical 3.2 you measured the biomass in plant material. Where did the carbon contained in this biomass originally come from?

b How is carbon passed from one trophic level to another?

Part 1: Respiration

All organisms use carbon in the form of carbohydrates to provide energy for all their needs. As they use these carbohydrates through respiration, the carbon is released as carbon dioxide.

In this part of the practical, you will show that carbon dioxide is produced by living organisms and describe how to test the product.

Method

1. Approximately half-fill a test tube with limewater.
2. Take a straw or piece of tubing and insert one end into the limewater in the test tube.
3. Gently blow through the straw or tube until you observe a change (Figure 3.2).

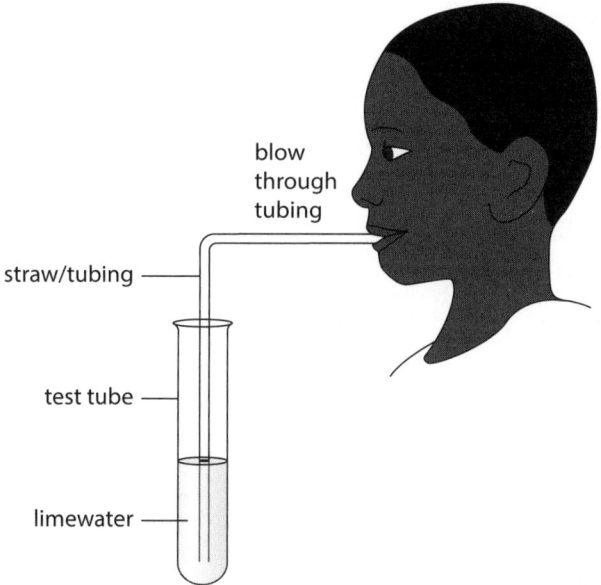

Figure 3.2: Diagram of apparatus to test the air exhaled when breathing out.

Results

Create and complete a table like Table 3.5.

Appearance of limewater before blowing into it	Appearance of limewater after blowing into it

Table 3.5: Observations when we blow exhaled air into limewater.

Part 2: Formation and chemical weathering of carbonate rocks

Chalk and limestone are the remains of marine organisms that lived and died millions of years ago, including calcareous phytoplankton and zooplankton. When the organisms were alive, they gained carbon through photosynthesis and feeding relationships in food chains. When they died, either they broke down through decay, or they could be trapped and fossilised in rocks.

In this part of the practical, you will show that these rocks contain carbon which can be released back into the atmosphere.

Method

1. Approximately half-fill a test tube with limewater.
2. Add 20 cm³ 1.0 mol dm⁻³ hydrochloric acid to a boiling tube.
3. Add a small piece of chalk or limestone into the acid in the boiling tube. Place the bung in the top containing the delivery tube.
4. Insert the other end of the delivery tube into the test tube of limewater. Bubble the gas produced through the limewater (as shown in Figure 3.3).

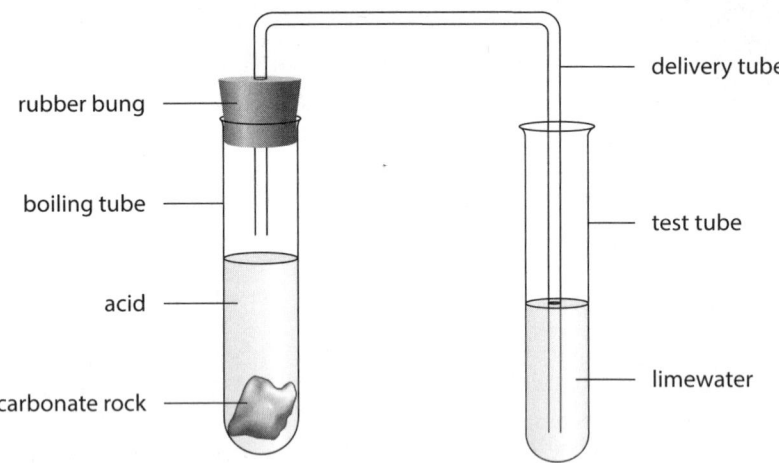

Figure 3.3: Diagram of apparatus to show the product from chemical weathering of carbonate rocks.

Results

Create and complete a table like Table 3.6.

Item	Observations and changes when chalk or limestone is added to the acid
size of chalk / limestone	
acid	
limewater	

Table 3.6: Observations when a carbonate rock is added to acid.

Part 3: Atmospheric dissolution into the oceans

The seas and oceans provide a reservoir of dissolved gases, including carbon dioxide which is vital both to the organisms living in the ocean and to the atmosphere in maintaining carbon dioxide levels. This part of the practical compares the solubility of carbon dioxide in fresh and salt water and the impact this has on acidity in the ocean.

Method

1. Add approx. 50 cm^3 of sea water to one beaker.
2. Add approx. 50 cm^3 of fresh water to another beaker.
3. Add a few drops of universal indicator solution to both beakers so that a distinct colour can be seen in both.
4. Gently blow into one of the beakers through a straw or tubing, timing how long it takes for the indicator to change to yellow. Record your results in a copy of Table 3.7.
5. Repeat with the other beaker, trying to blow with a similar intensity as you did for the first beaker. Record your results in a copy of Table 3.7.

> **TIP**
>
> Seawater is naturally weakly alkaline. Depending on the actual universal indicator used, you would expect the indicator to turn dark green.

Results

Water sample	Time taken for universal indicator solution to turn yellow / seconds
fresh water	
seawater	

Table 3.7: Observations of time taken for universal indicator to turn yellow in water of different salinities.

Part 4: Combustion

Fossil fuels contain carbon that has been locked away as fossilised remains of mostly marine organisms in rocks for millions of years. Many wax candles are made from paraffin wax which is produced from crude oil, a fossil fuel. Biofuels also contain carbon from organisms that have been alive much more recently, such as trees. This part of the practical is a demonstration which shows that burning both fossil fuels and biofuels releases carbon dioxide back into the atmosphere.

> **KEY WORD**
>
> **fossil fuel:** buried organic materials from dead plants and animals which have been converted into oil, coal or natural gas by exposure to heat and pressure in the Earth's crust

Method

1. Set up the apparatus as shown in Figure 3.4, with the boiling tube half full of limewater.
2. Turn on the suction pump so that a gentle stream of air moves through the apparatus, producing a stream of bubbles in the limewater.
3. Place the candle under the funnel. Observe and record any changes in the limewater in a table like Table 3.8.

4   Extinguish the candle and replace the limewater with some fresh solution.

5   Repeat the experiment by burning the small pieces of wood under the funnel, observe and record any changes in the limewater.

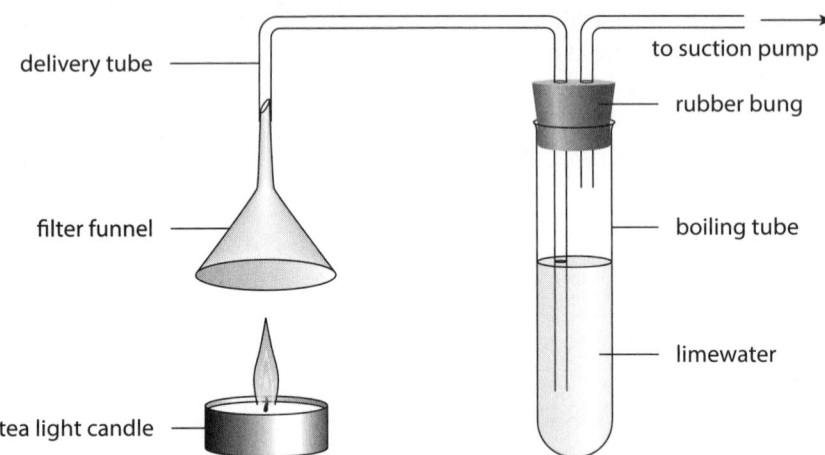

Figure 3.4: Diagram of how to set up apparatus to show the production of carbon dioxide from combustion of fuels.

Results

Fuel	Observations of the limewater when the fuel is burned
fossil fuel (paraffin wax)	
wood	

Table 3.8: Observations when different fuels are burned and the gases passed through limewater.

Evaluation and conclusions

c   Carbon dioxide is also present in the atmosphere. Describe a control experiment for Part 4 of the practical that would show that it is not just the carbon dioxide from the air causing the limewater to change.

d   How could the experiment in Part 1 be adapted to show that plants or other animals also respire?

e   Suggest how changes in the concentration of carbon dioxide in the atmosphere will affect the concentration of carbon dioxide in the oceans.

f   Suggest how deforestation might affect the concentration of carbon dioxide in the atmosphere. Explain your reasoning.

g   Another aspect of the carbon cycle is decomposition. Suggest a simple experiment that would show that decomposition also produces carbon dioxide.

> **TIP**
>
> You may want to consider using equipment from Part 4 of the practical.

Reflection

h   How have the activities you have carried out and observed in this practical helped your understanding of the carbon cycle?

Chapter 4
Classification and biodiversity

CHAPTER OUTLINE

In this chapter you will complete investigations on:
- 4.1 Constructing a dichotomous key
- 4.2 Using quadrats to estimate abundance of organisms
- 4.3 Estimating a population size using the mark–release–recapture method

Practical 4.1: Constructing a dichotomous key

Introduction
Living organisms have an incredible diversity, with many similarities and differences among them. Classification can be useful to help identify a species we are unfamiliar with when conducting fieldwork. This activity involves the construction of a dichotomous key, a tool that helps identify a specimen through a series of questions that separate species by their features. Although the activity has been described with gastropod molluscs, it can be done with any group of organisms, and it can be completed inside or out in the field.

EQUIPMENT

You will need:
- set of gastropod mollusc shells from different (for example, five) species
- hand lens.

Safety considerations
There are no significant safety issues associated with this practical investigation.

BEFORE YOU START

a. Why is it important that terms used for features are clearly identified?
b. Why are actual sizes and lengths on organisms often poor choices for identification keys?
c. How could you test your key to check that it works well?

Method

1. Study the group of shells you have been given carefully, using Figure 4.1 as a guide. Try to identify features or traits that you could use to separate the shells into groups.

> **TIP**
> Try to divide the group as equally as possible each time, rather than picking off unique features and separating each species one by one.

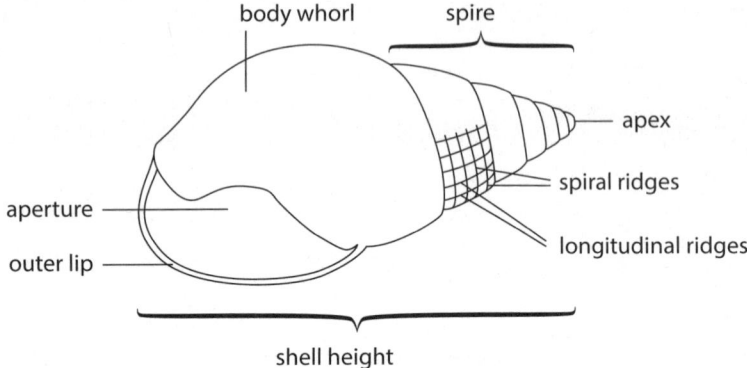

Figure 4.1: Labelled diagram of a gastropod mollusc shell.

2. Choose one of these features, record the feature in a dichotomous key (see Table 4.1 for how to create this) and split the group into two – those that have the feature (yes) and those that do not (no).

3. Repeat this process by choosing a feature to split each sub-group until each species is separated on its own. At this point, identify the species in the final column instead of noting which feature to move on to next.

4. Swap your key and collection of shells with another group. Can they use your key to correctly identify each species? If not, try to improve your key to make it easier for others to use.

5. An alternative method of presenting the key is to create a flow chart, with questions in boxes followed by arrows pointing to the next pair of questions, eventually identifying each species. Re-draw your key using this method.

> **TIP**
> To complete a copy of this table, write the feature you are using to separate the organisms into two groups in the main box. Complete the last column to direct which feature the person using the key should look at next. This has been done for you for feature 1. When a feature separates a single species, the final column should identify the name of the species.

Results

1. Feature:	Yes	Go to number 2.
	No	Go to number 3.
2. Feature:	Yes	
	No	
3. Feature:	Yes	
	No	
4. Feature:	Yes	
	No	
5. Feature:	Yes	
	No	

Table 4.1: Creating a dichotomous key.

4 Classification and biodiversity

Evaluation and conclusions

d Binomial nomenclature is an international convention for naming species. Suggest why is it important that keys are produced using the scientific names for species.

e In this activity you have used dichotomous keys to classify different species. Do you think that these keys could be used for classification of organisms at any level of classification (that is, to identify the kingdom, phylum, class, order, etc)?

Reflection

f Which method of creating a key do you think is the best to use? Do you think that this will be the same for everyone?

Practical 4.2: Using quadrats to estimate abundance of organisms

Introduction

A **habitat** will host **populations** of different species within a **community**. Many populations can be difficult to count exactly due to the difficulty of getting to all coastlines, or diving and the limitations on the time that surveys can be carried out due to the use of oxygen tanks. Instead we can use **random sampling** to estimate the population using **quadrats** placed randomly across the area to be surveyed. This activity helps you to understand the process of using quadrats. If you get an opportunity to go out on a field trip yourself, you can apply the same technique to collect data from observations in the field.

> **KEY WORDS**
>
> **habitat:** the natural environment where an organism lives
>
> **population:** all the individuals of the same species that live at the same place and time
>
> **community:** all the different populations interacting in one habitat at the same time
>
> **random sampling:** samples are taken at random places within the sample site
>
> **quadrat:** a square to mark an area, often divided into smaller squares; can be different sizes such as 1 metre × 1 metre or 0.5 metre × 0.5 metre

EQUIPMENT

You will need:

a method of generating random numbers (for example, search 'random number generator' on an internet search).

Safety considerations

There are no significant safety issues associated with this practical investigation.

BEFORE YOU START

a Why might it be useful to estimate the populations of different species in an area?

b A quadrat is a square that marks an area to be surveyed. The edges of the quadrat will sometimes include only part of an organism. What strategies could be used to count organisms that are only partly inside a quadrat?

c This practical uses a random number generator to select which quadrats to survey. Why do you think this is important?

d For sampling populations in a large area, why do you think it is important that at least ten quadrats should be surveyed?

CAMBRIDGE INTERNATIONAL AS & A LEVEL MARINE SCIENCE: WORKBOOK SECTION 2

Method

1. You are provided with an image (Figure 4.2) which has been divided into a grid of 0.5 metre × 0.5 metre quadrats. Each axis of the grid has been labelled with sequential numbers, 1–18 and 1–13, to create grid coordinates.

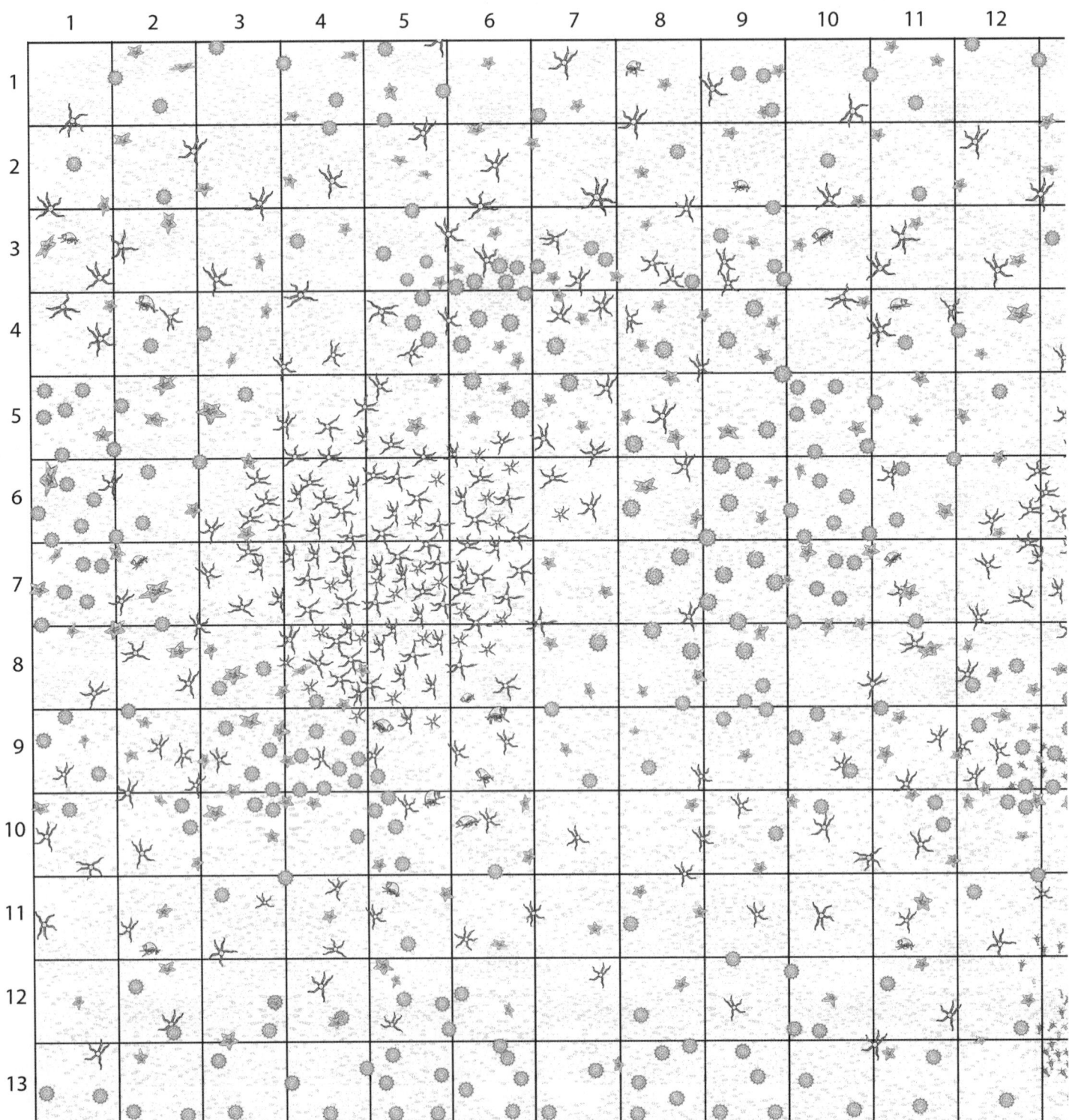

146

4 Classification and biodiversity

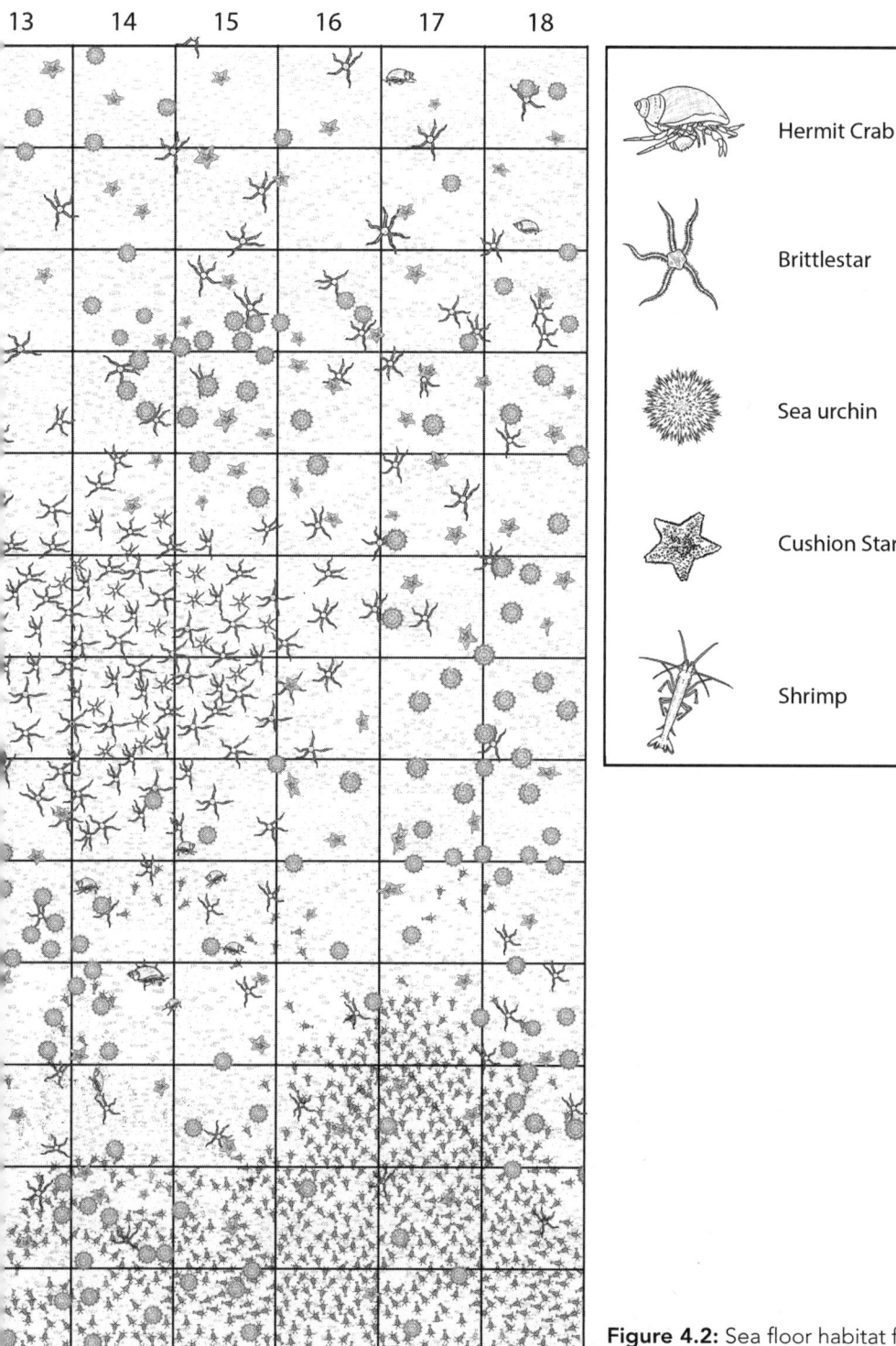

Figure 4.2: Sea floor habitat for Practical 4.2.

2 Use a random number generator to provide two numbers. These will indicate the first square or quadrat you count.

3 Count the number of individuals of one species in the square and record this in your table. Repeat this for each species present in the quadrat.

4 Repeat steps 2 and 3 for at least nine more squares, recording all your results in a table like Table 4.2. You will need 10 rows for the quadrat number along with rows for the total and mean.

TIP

If the number of individuals in a quadrat is difficult to count, you could divide it with a ruler into smaller sections (for example, quarters), so that you do not lose count as easily.

Results

Quadrat no.	Brittlestar	Cushion starfish	Sea urchin	Hermit crab	Shrimp
1					
...					
10					
Total					
Mean					
Estimated total population					

Table 4.2: Results table to collect and process data for quadrat practical.

Evaluation and conclusions

e Calculate the total for each species in the ten selected quadrats, and the **mean** number of each species in this habitat. Record this information in the boxes in your table to an appropriate number of significant figures.

f Use the average to estimate the total population for each species in the total area of the habitat shown. Record this information in your table.

KEY WORD

mean: the sum of a number of items of data, divided by the total number of items of data added

TIP

Calculate the total area in the habitat (each square is 0.5 metre × 0.5 metre), then multiply the mean calculated in **e** by this area:

total no. of individuals in habitat sampled =

$$\frac{\text{total area of habitat}}{\text{area of 1 quadrat}} \times \text{mean no. of that species}$$

g Compare your results to those calculated by other groups. How similar are they? If they are very different, suggest why this may have occurred.

Reflection

h How could you improve the validity of your results?

i If you were carrying out a survey like this to monitor changes in populations over time, how would you try to ensure that your results were comparable over extended periods of time?

j This method can be used for real-life population surveys. How well do you think this activity represents using quadrats in real life? What do you think would be more difficult when carrying this out for real?

Practical 4.3: Estimating a population size using the mark–release–recapture method

Introduction

The mark–release–recapture method can be used to estimate the population size of an organism (for example mobile invertebrates on a rocky shore). To do this you will apply the Lincoln index, a formula using the number of organisms counted on two separate occasions. This can be carried out for organisms that are both numerous in the chosen location and able to move around between sampling. In this activity you will model the use of this technique to estimate the number of dry pasta shells in a bag of pasta. Depending on your location and circumstances, you may be able to follow this up with an investigation into a population of small invertebrates near you.

EQUIPMENT

You will need:

- packet of dried pasta shells (or other small items such as marbles)
- large bag or box to place the pasta into
- permanent marker.

Safety considerations

There are no significant safety issues associated with this practical investigation.

BEFORE YOU START

a It is relatively easy to count all the pieces of pasta in a bag. Why might it be more difficult to count organisms in their habitat?

b Some of the pasta shells might be broken or damaged. How will you deal with these when counting?

TIP

The pasta shells represent the animals we are estimating the population of. The bag or box represents their habitat.

Method

1. Empty the pasta into a box or large bag (the 'habitat').
2. Remove a handful or two of shells (the first sample); count and mark these. Record the number of shells marked in a table like Table 4.4 and replace the marked shells into the 'habitat'.
3. Shake / mix all the shells in the 'habitat'.
4. Remove a similar number of shells (one or two handfuls). Count the number that are marked and unmarked (second sample) and record in a results table like Table 4.4.

Results

Number of shells marked in first sample	Tally:	Total number marked:	
Number of shells in second sample	Marked shells	Unmarked shells	
	Tally:	Tally:	
	Total number marked:	Total number unmarked:	

Table 4.4: Results table for mark–release–recapture practical.

Evaluation and conclusions

c Calculate the estimated number of 'animals' (pasta shells) in the whole 'population' (packet), using this formula:

$$n = \frac{n_1 \times n_2}{m_2}$$

in which:

N = (estimated) population size

n_1 = number of individuals caught, marked and released in the first sample

n_2 = number of individuals caught in the second sample (both marked and unmarked)

m_2 = number of marked individuals recaptured in the second sample.

d Compare your results to those from another group. How similar are your results?

e How could you make your results more reliable?

f The **biodiversity** of species depends on both the number of different species in a habitat and the populations of each species. Suggest why the mark–release–recapture method is a useful tool for measuring species biodiversity, and why this is not suitable for all species.

> **KEY WORD**
>
> **biodiversity:** a measure of the species, genetic and ecosystem diversity of different species

Reflection

g Think about the limitations of this classroom activity. How do you think this compares to a real fieldwork exercise?

Chapter 5
Examples of marine ecosystems

CHAPTER OUTLINE

In this chapter you will complete investigations on:
- 5.1 Drawing an animal found on a sandy shore
- 5.2 Planning an investigation into the effect of light intensity on coral growth
- 5.3 Distribution of organisms on a rocky shore

Practical 5.1: Drawing an animal found on a sandy shore

Introduction

Sandy shores are **unstable habitats** that often initially appear to be deserted. On closer inspection the sand can be home to many organisms that have adapted to the moving **substrate**.

This task practises your drawing skills while identifying features that help the organisms to adapt to their habitat.

EQUIPMENT

You will need:
- a pencil, HB or 2H
- pencil sharpener
- ruler
- eraser
- calculator.

KEY WORDS

unstable habitat: a habitat that has moving or shifting substrate making it difficult to attach to

substrate: the material that makes up the sea floor, such as rocks, sand, silt, etc.

Safety considerations

There are no significant safety issues associated with this practical investigation.

BEFORE YOU START

a Biological drawings are not intended to be like photographs. Suggest what aspects are most important when drawing specimens.

b What type of features would you expect to find in animals that live on a sandy shoreline?

Method

You are provided with an image (Figure 5.1) showing a blow lugworm (*Arenicola marina*), an annelid (segmented worm), burrowed in sand with a 'cast' of sand excreted by the lugworm at the surface. (This is a common visible sign of the presence of lugworms on the beach.)

Figure 5.1: Blow lugworm, *Arenicola marina*.

1 Draw a *larger* copy of the lugworm and its burrow and cast on plain paper.

> **TIP**
>
> Always draw with a single continuous line. Never draw 'fuzzy' or multiple lines for the same line. Use a pencil so you can erase the line and redraw if necessary. You are not expected to be a great artist. You do need to show approximately the correct *shape* and *proportions* of different parts of the organism. Make sure you include relevant *details* that are clearly visible in the photograph.

2 The lugworm has several bushy protrusions along its body. These are the **gills**. Label this on your diagram.

3 The lugworm also has bristles along its body. Label a bristle on your diagram.

4 Annelids consist of three regions or sections – the head (anterior section), thoracic section and the tail (posterior) section. The bristles and gills are located on segments that make up the thoracic region of the lugworm. Label the section of the worm that is the thoracic region.

5 The cast is excreted from the tail section of the lugworm. Label this section and the head section on your diagram.

> **KEY WORD**
>
> **gills:** the gaseous exchange surfaces of fish

Evaluation and conclusions

c How many segments on the lugworm have gills? Suggest why the lugworm has multiple gills exposed along its body.

d Suggest the purpose of the bristles along the body.

e The lugworm in the image is 20 cm long. Assume the 'U' shape on the image has a diameter corresponding to a circle with 20 cm circumference. Calculate the magnification of *your drawing*.

f At low tide, many birds can often be seen feeding on lugworms. Suggest how the birds locate the lugworms and how the birds might be adapted to obtain the lugworms from their burrows in the sand.

> **TIP**
>
> The circumference and diameter of a circle are related by this formula:
>
> $$\pi = \frac{\text{circumference}}{\text{diameter}}$$

> **TIP**
>
> $$\text{magnification} = \frac{\text{image size}}{\text{actual size}}$$

Reflection

g Compare your drawing to another student's drawing. Which aspects did you draw better, and which aspects did they draw better?

Practical 5.2: Planning an investigation into the effect of light intensity on coral growth

Introduction

Coral reefs are built by tiny coral polyps that live in close proximity to photosynthetic **zooxanthellae**. Therefore, the growth of corals depends on the intensity of the light they receive, as well as a number of other factors. In this investigation you will plan a laboratory-based investigation into how light intensity affects the rate of growth in corals.

> **KEY WORD**
>
> **zooxanthellae:** symbiotic, photosynthetic dinoflagellates living within the tissues of many invertebrates

> **BEFORE YOU START**
>
> a What variables, apart from light intensity, could affect the growth of coral polyps?
>
> b Write a hypothesis that you are planning to investigate.
>
> c What measurements will you need to collect to determine if your hypothesis is correct?

Planning

Variables

1 To produce valid data, you need to identify the variables that you will change (independent) and measure (dependent). To calculate the rate of growth, you need measurements of both time and growth of coral.

   i   What is the independent variable?

   ii Choose five or six different values for the independent variable.

   iii How many times will you repeat each measurement?

   iv How will you change the independent variable in your experiment?

   v   What is the dependent variable?

   vi How will you measure the dependent variable?

2 To produce valid data, you must try to control other variables that may affect your results. These are called *control variables*. Some of these may vary when you change your independent variable, so you should plan how to minimise how much they will change. Copy and complete Table 5.1.

Variable	How to keep it constant

Table 5.1: Control variables for light intensity investigation.

Equipment

3 List the equipment you will need to complete your investigation in a table like Table 5.2.

Equipment	Quantity

Table 5.2: Equipment for light intensity investigation.

Safety precautions

4 What hazards might your method involve? How could you reduce the risk from these hazards? Copy and complete Table 5.3.

Hazard	Steps taken to reduce the risk from this hazard

Table 5.3: Safety precautions for light intensity investigation.

5 Corals are living organisms. What steps can you take to ensure that the investigation is ethical?

Method

6   Write a full step-by-step method for how to carry out the experiment. Include a diagram if this helps explain how to set up apparatus. Include all the practical details you need to carry out your experiment, including how to change the independent variable, how to measure the dependent variable and how to keep the control variables the same.

Results

7   Draw a results table to record the data you plan to collect. Make sure to include spaces for all the measurements you plan to take, including repeats and means. Ensure that all appropriate units are included in headings.

8   What do you predict the results of your investigation would be? How could you use your data to draw conclusions?

9   What type of graph would you use to present your results? Sketch the axes that you would use to plot this graph. Explain why this type of graph is most suitable.

Reflection

10 Swap plans with another student and read their method. How could their plan be improved? Do you think you would be able to complete the investigation without any help from the writer? Suggest improvements on each other's methods, to give each other feedback on how to improve your method.

Practical 5.3: Distribution of organisms on a rocky shore

Introduction

In a habitat where conditions are fairly consistent across the habitat, you can use random sampling to measure the abundance of different species across the habitat. Many shorelines have gradual changes in conditions due to the zones formed from the changing tides. In these habitats it is useful to know about the **distribution** of organisms from the top of the shore to the low-tide mark as well as their abundance, so a more **systematic sampling** method called a **transect**, along with quadrats, can provide this additional information. This activity helps you to understand the use of a systematic sampling technique and to compare it with the random sampling that you used in Practical 4.2. If you get an opportunity to go out on a field trip, you can apply the same technique to collect data from observations in the field.

> **KEY WORDS**
>
> **distribution:** the variation of a population across an area
>
> **systematic sampling:** samples are taken at fixed intervals along the transect
>
> **transect:** a rope or tape marked at regular intervals that sets standard distances for study of the distribution of marine organisms

EQUIPMENT

You will need:

- a small transparent plastic square to use as a quadrat (for example, 3 cm × 3 cm)
- a length of string to mark a transect line.

Safety considerations

There are no significant safety issues associated with this practical investigation.

Figure 5.2: Rocky shoreline to investigate distribution of organisms.

5 Examples of marine ecosystems

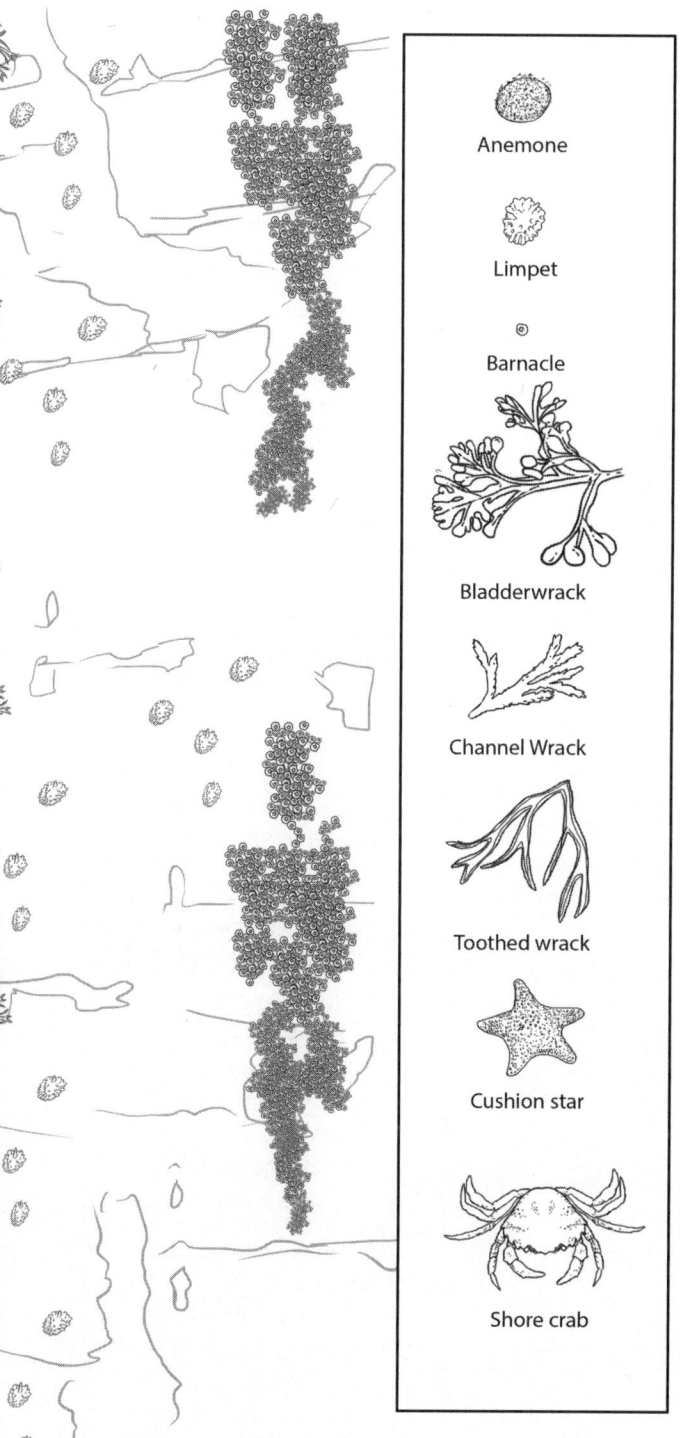

157

CAMBRIDGE INTERNATIONAL AS & A LEVEL MARINE SCIENCE: WORKBOOK SECTION 2

> **BEFORE YOU START**
>
> In this activity you will be using a transect to measure changes in the abundance of different species from the low-tide mark to the top of the shore. Look at Figure 5.3 and read the following information about different types of transect.
>
>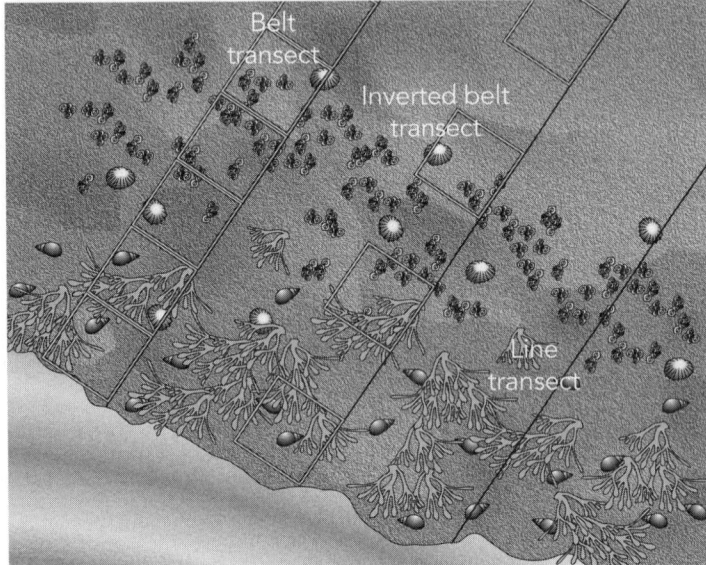
>
> **Figure 5.3:** How to carry out different types of transect.
>
> - **Line transect** – record every species touching the line along its whole length.
> - **Continuous belt transect** – place your quadrat next to the line and record the abundance of each species in the quadrat (by counting or percentage cover, as appropriate for each species). Repeat these readings by moving the quadrat up the line without leaving any gaps between.
> - **Interrupted belt transect** – record the abundance along the transect using a quadrat but leave regular spaces between areas sampled with the quadrat.
>
> **a** Look at the site to be surveyed (Figure 5.3) and the size of the quadrat square you will be using. Suggest why a continuous belt transect is the most suitable technique for this shoreline.
>
> **b** Look at the different species present on Figure 5.3. For each species, suggest whether it would be best to count each individual or estimate the percentage of the quadrat that the species covers.

TIP

A line transect is best for very long transects and when you do not have a lot of time to record your results.

KEY WORDS

continuous belt transect: use of a quadrat to collect population information along a transect without any gaps

interrupted belt transect: use of a quadrat to collect population information along a transect with regular gaps between samples

Method

1. Using the string, mark a transect line on the shore in Figure 5.3 that you will investigate from the sea to the upper zone.

2. Start at the low tide mark (on the left of Figure 5.3). Place your quadrat next to the line so that it just touches the sea. Count (or estimate the percentage cover) for each species present in the quadrat. Record your results in a results table (see question **c**).

3   Move the quadrat along the transect line away from the sea so that it touches where it was previously. This can be done by flipping the transect on one side to move it up the transect line, as shown in Figure 5.4.

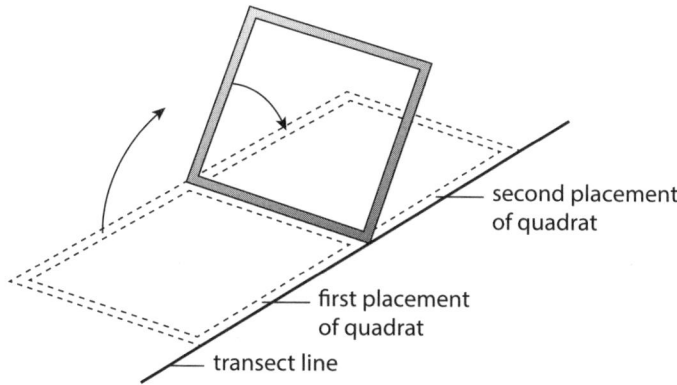

Figure 5.4: How to move a quadrat along a belt transect.

4   Count (or estimate the percentage cover) for each species present in this second quadrat. Record your results in the results table (see question **c**).

5   Repeat steps 3 and 4 until you have recorded the distribution of each species all along the transect.

> **TIP**
>
> A continuous belt transect is best for very short transects and when you have lots of time to record your results.

> **TIP**
>
> An interrupted belt transect collects a representative sample of results and can be used for longer distances where you want to indicate the abundance as well as where different species are present.

Results

c   Create a results table in which to record all your results for each species along the full length of the transect. Make sure to include spaces to record the distance along the shore and the abundance of each species for each distance.

A **kite graph** is a type of graph showing the change in distribution of organisms along a transect. An example is shown in Figure 5.5.

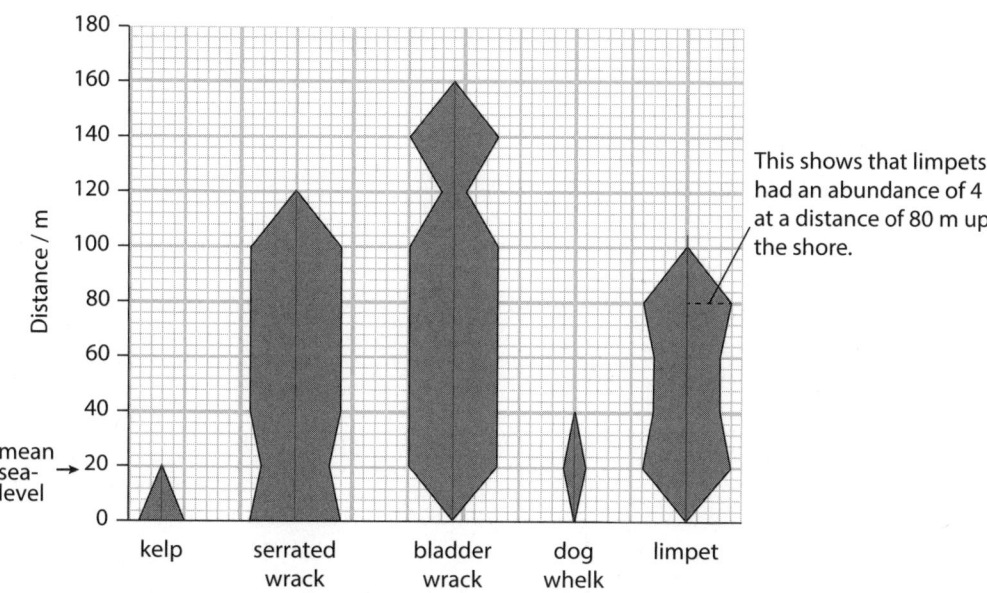

Figure 5.5: Kite graph.

> **KEY WORD**
>
> **kite graph:** a graph of the distribution and abundance of organisms in the littoral zone that allows zonation patterns to be easily seen

When drawing a kite diagram:

- Use graph paper.
- Decide which way round you want your 'kites' to run (vertically or horizontally). The axes can be either way round.
- Construct the distance scale to the full length of the transect you have used.
- List the species along the other axis, spacing them evenly apart (similarly to how you would for a bar chart). Plan how much space you need for each species. Read the next bullet point before doing this.
- Each 'kite' represents the abundance of a species at different distances along the transect. Use a ruler to draw a faint pencil line vertically upwards, or horizontally, from the centre position of each species (this represents the transect line). Then use your results to plot points representing the population of that species at each measured position, *on both sides* of this central line.
- Join your points with ruled straight lines, to make symmetrical shapes centred on the pencil line you have drawn for each species.

d Draw a kite diagram to show your results. Label each 'kite' clearly with the species it represents. Mark the distances on the distance scale, including units. Plot all points accurately and join the points with ruled lines.

Evaluation and conclusions

e Select two species that you have plotted that have different distributions along the transect. Describe the distribution of each of these species.

f Suggest why the species measured have different distributions along the transect.

g Use your results from the investigation to suggest how to identify the locations of the lower shore, middle shore, upper shore and splash zone for this shoreline.

h Choose a different type of transect from the one you used to suggest how your results might differ if they were collected using a different type of transect from the method you used.

> **TIP**
>
> Your suggestion should consider how the conditions they are exposed to along the transect will change during each tidal cycle, and how the organisms can be adapted to cope with these regular changes.

Reflection

i Discuss how well the method used represents a real field-based investigation. What additional challenges might you expect to face if you were carrying this out on a real shoreline?

Chapter 6
Physiology of marine organisms

CHAPTER OUTLINE

In this chapter you will complete investigations on:

- 6.1 Observing, drawing and comparing the structures of respiratory systems
- 6.2 Investigating the effect of salinity on brine shrimp
- 6.3 The effect of salt solution on eggs

Practical 6.1: Observing, drawing and comparing the structures of respiratory systems

Introduction

Animals with larger bodies require **ventilation** and **circulatory systems** to help **gas exchange** take place efficiently for all cells. In this activity you will examine and compare the gas exchange systems in fish and mammals, and then observe and draw these structures.

EQUIPMENT

You will need:

For the demonstration

- lungs and trachea from a mammal (for example, a lamb – one per class)
- head of a large fish (for example, a salmon – one per class)
- dissection board(s)
- dissecting knife / scalpel
- dissecting scissors
- forceps.

For the practical

- light microscope
- prepared section of TS alveoli
- prepared section of TS gill lamellae
- pencil (HB or 2H)
- eraser
- pencil sharpener
- ruler.

KEY WORDS

ventilation: the exchange of gases (oxygen and carbon dioxide) into and out of organisms; the term also describes a method of achieving this, (for example, breathing in mammals and ram or pumped ventilation in fish)

circulatory system: an organ system comprising the heart, blood vessels and the blood; its role is to carry useful substances around the body to all cells and move waste products away from the cells to be removed

gas exchange: the uptake of oxygen and release of carbon dioxide by cells other surfaces

Safety considerations

Follow all safety instructions given by your teacher.

> CAMBRIDGE INTERNATIONAL AS & A LEVEL MARINE SCIENCE: WORKBOOK SECTION 2

> **BEFORE YOU START**
>
> **a** Why is it important that we observe the respiratory systems at a microscopic level in addition to the larger structures to understand how they aid gas exchange?
>
> **b** What safety precautions should be taken when handling fresh biological materials such as lungs and fish?

Part 1: Observing and comparing the dissection of the respiratory systems of a mammal and a fish

Method

1 Your teacher will show you the overall anatomy of the respiratory system of a land-based mammal such as a lamb. During the demonstration draw diagrams and make notes on features and adaptations that your teacher identifies (see Results section).

2 Your teacher will show you the overall anatomy relating to the respiratory system of a bony fish. During the demonstration draw diagrams and make notes on features and adaptations that your teacher identifies (see Results section).

Results

c During the demonstration of the dissection of the lungs, draw diagrams and make notes on features and adaptations that your teacher identifies:

- overall appearance of the lungs: colour, texture, elasticity
- trachea: shape and function of the cartilage, diameter
- bronchi: shape and function of the cartilage, diameter
- terminal bronchioles: shape and function, absence of cartilage, diameter.

d During the demonstration of the dissection of a bony fish draw diagrams and make notes on features and adaptations that your teacher identifies, including:

- overall appearance of the gills: colour, texture, location relative to operculum and mouth
- gill arches: number of, shape, blood supply
- gill filaments: shape, position, appearance.

Part 2: Observation and plan drawing of the structure of alveoli

Method

1 Set up a light microscope.
2 Take the prepared slide of TS alveolus and view using a high-power objective lens.
3 The alveoli will be visible as large spaces separated by thin layers of tissue. You might also observe blood vessels and bronchioles.
4 Draw and label a plan diagram (not showing individual cells). See Results section.
5 Take a prepared slide of TS gill lamellae and view using a high-power objective lens.
6 The lamellae should be visible as long thin protrusions from the main gill filament.
7 Draw and label a plan diagram of the gill lamellae. See Results section.

> **TIP**
> You will need to focus initially using a low-power objective lens.

Results

Draw and label a plan diagram (not showing individual cells) of the alveoli to show:

- two or three alveoli
- layers of tissue (squamous epithelium) separating separate alveoli
- blood vessels or bronchiole (if present).

Draw and label a plan diagram of the gill lamellae to show:

- two or three lamellae
- the section of gill filament these lamellae are attached to.

Evaluation and conclusions

e How does the microscopic structure of the lungs help increase the rate of diffusion for gas exchange?
f How does the microscopic structure of the gills help increase the rate of diffusion for gas exchange?
g Describe and compare the ventilation of the lungs and the gills. How does each type of organ supply oxygen for gas exchange? How do both mammals and fish adapt to increase the supply of oxygen?
h Once gas exchange has occurred, how is oxygen transported to cells?
i The lungs that you observed were from a land-based (terrestrial) animal. Suggest adaptations that might be needed to this ventilation system for marine mammals, such as whales, that need to stay underwater for extended periods of time.

Reflection

j How has this activity developed your understanding and awareness of how larger organisms get oxygen to all their cells and remove the carbon dioxide produced?

Practical 6.2: Planning an investigation into the effect of salinity on brine shrimp

Introduction

Brine shrimp (*Artemia* sp.) are aquatic crustaceans which have rapid lifecycles, making them popular for use in **aquaculture** as a food source for other species. This rapid lifecycle along with their tolerance for a wide range of conditions make them useful organisms to investigate. In this practical you will plan and carry out an investigation into the ideal conditions for brine shrimp to hatch and the extent to which they are **stenohaline** or **euryhaline**.

> **BEFORE YOU START**
>
> **a** In this investigation you are going to look at the effect of salinity on the hatch rate of the brine shrimp. Write a hypothesis that links the variables, making clear what you expect the salinity to do to the hatch rate of the brine shrimp.
>
> **b** Support your hypothesis with a reason, using your knowledge of salinity, for the movement of substances and the ability of organisms to tolerate a range of different salinities.

> **KEY WORDS**
>
> **aquaculture:** the rearing of aquatic animals and plants for human consumption or use
>
> **stenohaline:** organisms that cannot tolerate wide changes in the salinity of water
>
> **euryhaline:** organisms that can tolerate wide changes in the salinity of water

Planning

1. The independent variable is the salinity. Suggest a suitable range of salinities to use in your investigation. Suggest what intermediate values you will use. You should plan to use at least five different salinities spread across your chosen range.

2. How many times should you repeat your readings to get reliable results?

3. How, and when, will you measure your independent variable (the hatch rate of the brine shrimp)?

4. What control variables do you need to consider? How can you ensure these are kept as similar as possible?

5. What equipment will you need to complete your investigation?

6. What hazards will your investigation involve? What steps can you take to reduce the risks from these hazards?

7. Brine shrimp are living organisms. Describe precautions you should take to ensure the experiment is carried out ethically.

Method

8. Use your answers to questions 1–7 above to write a clear, step-by-step method for your investigation. Use bullet points to help structure your answer to make it easy for someone else to follow your instructions.

Results

9. Draw a results table to record the data you plan to collect. Make sure you include spaces for all the measurements you plan to take, including repeats and means. Ensure that all appropriate units are included in headings.
10. Check your plan with your teacher. Once it has been agreed, carry out your investigation and complete the results in your table.

Evaluation and conclusions

c. Plot a suitable graph of your results, making sure to label the axes clearly with appropriate units and to plot all points accurately.
d. What did you find is the best salinity for brine shrimps? Did you get the same results as other groups?

Reflection

e. Look back at your method. How could sources of error could have caused inaccuracies in your results?
f. How could you improve your method to improve your confidence in the results?

Practical 6.3: The effect of salt solution on eggs

Introduction

The membranes of eggs are partially permeable, allowing water molecules to pass through but not sodium ions or chlorine ions. In this investigation, you will cover hen's eggs in sodium chloride solutions of different concentrations. You will investigate the effect of the different solutions on the mass of the eggs over time.

EQUIPMENT

You will need:

Day 1

- five hen's eggs
- a very large beaker or other container to hold all five eggs
- a large spoon or other implement for lowering the eggs into the acid (or gloves)
- enough 1.5 mol dm^{-3} hydrochloric acid to cover the eggs in the beaker.

Day 2

- a large spoon or other implement for removing the eggs from the acid
- about 400 cm^3 20% sodium chloride solution
- 2 × 200 cm^3 measuring cylinder
- dropping pipettes
- a timer
- 5 × 400 cm^3 beakers
- paper towels
- electronic balance
- distilled water.

Safety considerations

Hydrochloric acid is irritant at the concentration used. Wear eye protection and avoid splashing onto skin. Wash with plenty of cold water if any comes in to contact with the skin.

> **BEFORE YOU START**
>
> a   Explain the difference between **diffusion** and osmosis.
>
> b   What is meant by the term **'water potential'**? Which would have a higher water potential – ocean water or fresh water?
>
> c   Suggest reasons why a stronger concentration of acid is not used to remove the hard shells.

> **KEY WORDS**
>
> **diffusion:** the random movement of particles (or molecules) from a higher concentration to a lower concentration (down a concentration gradient); it is a passive process, not requiring the input of energy
>
> **water potential:** the potential energy of water in a solution compared to pure water; water will move by osmosis from a higher water potential to a lower water potential

Method

Day 1: Removing the hard shells from the eggs

The hard shells of birds' eggs contain calcium carbonate. This can be removed overnight by reacting the shells with acid to form carbon dioxide and water, and this will leave the partially permeable membrane underneath undamaged.

1   Place five hen's eggs into the large beaker.

2   Pour enough hydrochloric acid into the container to completely cover all the eggs.

3   Place in a safe place and leave overnight. (Mark on the beaker that it contains 1.5 mol dm^{-3} hydrochloric acid in case of any accident while you are not present.)

Day 2: Setting up the experiment

Make up solutions of sodium chloride of five different concentrations.

4   You are provided with a 20% sodium chloride solution. Use Table 6.1 to make up a range of five different concentrations of sodium chloride solution. You will need to make 200 cm^3 of each solution.

Final concentration of solution / %	Volume of sodium chloride solution added / cm^3	Volume of distilled water added / cm^3
0	0	200
5	50	150
10	100	100
15	150	50
20	200	0

Table 6.1: Diluting sodium chloride solution to different concentrations.

5   Carefully remove the hen's eggs, one at a time, from the hydrochloric acid. Wash each egg in water, very gently, and dry using a paper towel. Take care not to break the membrane that surrounds the egg.

6   Measure the mass of each egg in turn and record these measurements in a results table like Table 6.2. Place each egg in a labelled beaker.

7 Start a stopwatch. Pour the different concentrations of salt solution over each egg, making sure that the egg is completely covered.

8 After the eggs have been in their solutions for 30 minutes, gently remove each egg and dry it. Measure its mass and record this value in your results table.

9 Calculate the percentage change for each egg.

> **TIP**
>
>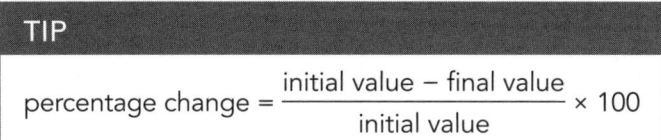
> $$\text{percentage change} = \frac{\text{initial value} - \text{final value}}{\text{initial value}} \times 100$$

> **TIP**
>
> It might be easier to record the final masses if you plan to pour the solutions over the eggs at different times (for example, one minute apart from each other).

Results

Concentration of sodium chloride solution / %	Initial mass of egg / g	Final mass of egg / g	Percentage change in mass of egg / %
0			
5			
10			
15			
20			

Table 6.2: Table to record changes to mass of eggs in different salt solutions.

Evaluation and conclusions

d Plot a graph of concentration of sodium chloride solution against the percentage change in the mass of egg.

e Why is it important to use percentage change in mass rather than actual change in mass of each egg?

f Describe the effect of sodium chloride solutions on the percentage change in mass of the eggs.

g Would your results suggest that hen's eggs are **osmoconformers** or **osmoregulators**? Why?

h Suggest why, in terms of osmosis, salmon return to fresh-water streams to lay their eggs.

i Birds' eggs must be exposed to air so that the developing chick can obtain oxygen and lose carbon dioxide by diffusion through the egg membranes. To what extent do you think this experiment can be used to confirm that the cell membranes in marine species act in the same way? Suggest how the investigation might be adapted to confirm your results using eggs from a marine species.

Reflection

j Having finished the practical, how do you think that you could have improved the reliability of your results?

> **KEY WORDS**
>
> **osmoconformer:** organisms that have an internal body fluid salinity that it is the same salinity as external water
>
> **osmoregulator:** organisms that regulate the internal salinity of their body fluids within a narrow range

Chapter 7
Energy

CHAPTER OUTLINE

In this chapter you will complete investigations on:
- 7.1 Identification and separation of photosynthetic pigments using paper chromatography
- 7.2 Data analysis into limiting factors for photosynthesis
- 7.3 Gas exchange in an aquatic producer

Practical 7.1: Identification and separation of photosynthetic pigments using paper chromatography

Introduction

Producers contain several primary and accessory pigments that combine to absorb different wavelengths of light for photosynthesis. Paper **chromatography** can be used to separate mixtures of solutes based on their different solubility in the solvent used. This practical uses chromatography to compare the photosynthetic pigments used by seaweeds at different heights on the shore.

KEY WORD

chromatography: a technique used to separate substances by their solubility

EQUIPMENT

You will need:

- samples of seaweeds from different heights on the shore, to include green, red and brown species
- pestle and mortar
- a small amount of washed, dried sand
- small beaker (for example, 100 cm^3)
- large beaker (for example, 250 cm^3) – to cover the small beaker
- 3 × very small beakers (for example, 10 cm^3)
- chromatography paper to fit in the beaker
- drawing pin or paper clip and a short pencil / wooden dowel to support the chromatography paper
- 3 × glass capillary tubes
- propanone (approx. 30 cm^3)
- chromatography solvent – 9 parts petroleum ether (80–100 °C), 1 part propanone (approx. 20 cm^3)
- HB or 2H pencil
- pipette
- filter funnel and muslin cloth
- centrifuge and centrifuge tubes (if available).

Safety considerations

Propanone and petroleum ether are both flammable and harmful. Take care not to breathe them in. Take care with glass capillary tubes, which are easily broken. Wear eye protection throughout this practical.

7 Energy

> **BEFORE YOU START**
>
> a Examine the samples of seaweeds available. Do any of them have air bladders? If so, what colour are these specimens and what height on the shore are they found?
>
> b Which wavelengths / colours of light are absorbed nearer to the surface in bodies of water?
>
> c What are the advantages and disadvantages for seaweeds in being able to absorb a different range of wavelengths of light?

Part 1: Extraction of pigments

Method

1 Use scissors to cut up approximately 20 g of fresh seaweed blades. Only use the blades. Avoid using the stipe or other parts.

2 Place the cut pieces into the mortar, add a small amount of sand and approximately 10 cm³ of propanone.

3 Use the pestle to grind the pieces (see Figure 7.1) until the propanone has turned a dark green or brown, containing a concentrated extract. If you need to add more propanone, only add a small amount (1–2 cm³) or it may make the extract too dilute to work.

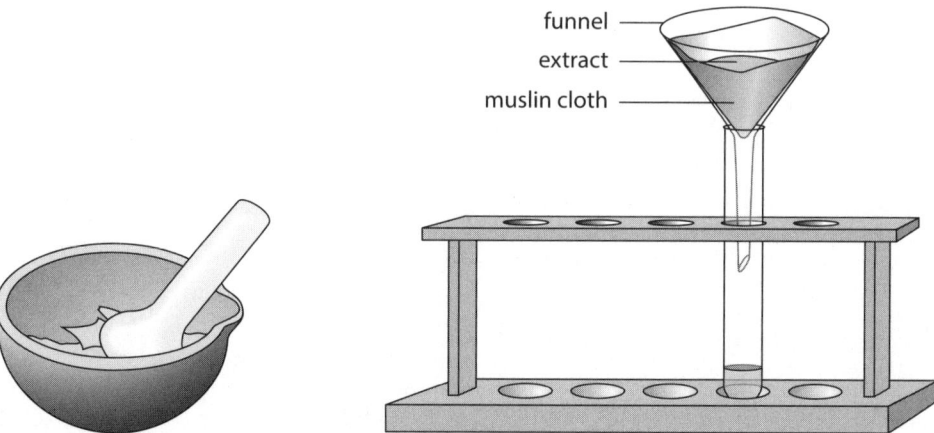

Figure 7.1: Extracting the pigments from seaweed.

4 Carefully pour the extract through a muslin cloth and funnel into a very small beaker. Use the pestle to squeeze as much liquid extract from the seaweed as possible.

5 If a centrifuge is available, place the extract into a centrifuge tube and spin for 2 minutes. If a centrifuge is not available, leave any solids to drop for at least 5 minutes.

6 Repeat steps 1–5 with the other two types of seaweed, to produce a total of three extracts, labelled 'g' (green), 'b' (brown) and 'r' (red).

Part 2: Chromatography

Method

7　Cut a piece of chromatography paper to fit in the 100 cm³ beaker so that it does not touch the sides of the beaker. Attach the top of the paper to a pencil using a drawing pin or paper clip so that the bottom of the paper does not quite touch the bottom of the beaker.

8　Use a pencil to lightly draw a horizontal line approximately 15 mm from the bottom of the paper. Use your pencil to mark three equally spaced 'x'' marks on the line, label these 'g', 'b' and 'r' to represent the extracts from the green, brown and red seaweeds.

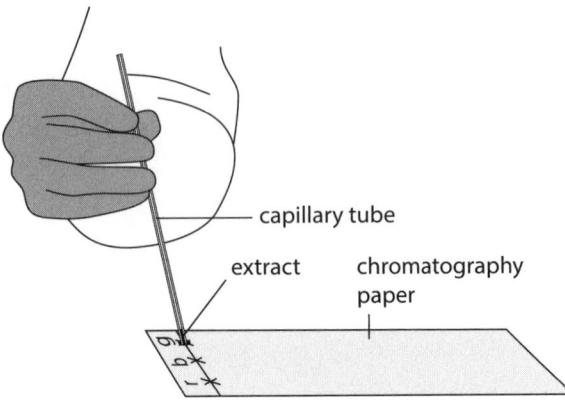

Figure 7.2: Applying the pigments to the chromatography paper.

9　Place a capillary tube into the first extract from the green seaweed. Some of the extract will be drawn up into the tube. Carefully and lightly touch the first 'x' marked 'g' so that a small spot of extract marks the middle of the 'x' (see Figure 7.2). Gently wave the paper in the air so the solvent can evaporate. The aim is to build up a concentrated spot of extracted pigments on the 'x' by allowing the solvent to evaporate between each spot so it does not spread out too much. Repeat this at least ten times to make the spot dark and concentrated.

10　Repeat step 9 for each of the other extracts on the 'x's' marked 'b' and 'r'.

11　Add the chromatography solvent to the beaker to a depth of no more than 10 mm. It is important that the solvent remains below the line and the spots of extracts, but that it can reach the bottom of the chromatography paper.

12　Suspend the paper in the beaker and cover with the larger beaker. This allows the solvent to create a more saturated 'atmosphere' inside the beaker and produces better results faster.

13　This could take 1–2 hours, so do not disturb the equipment while obtaining the results.

14　When the solvent has moved to within 10 mm of the top, remove the paper and quickly mark (in pencil) where the solvent reached (this is the solvent front).

15　Allow the chromatogram to dry. Gently circle each pigment that separates and mark the centre of each spot with a light 'x'. Measure the height that each pigment has moved up the paper and record your results in a table like Table 7.1.

Results

Distance moved by each pigment / mm	Extract 'g'	Extract 'b'	Extract 'r'
pigment 1			
pigment 2			
pigment 3			
pigment 4			
pigment 5			

Table 7.1: Results for separating photosynthetic pigments.

Evaluation and conclusions

d How many pigments were in each of the extracts?

e Calculate the R_f value for each pigment and record in a table like Table 7.2.

TIP	TIP
$R_f = \dfrac{\text{distance moved by pigment}}{\text{distance moved by solvent}}$	All R_f values have a value between 0 and 1. If your answer is below 0 or greater than 1, then check your calculation.

	Extract 'g'		Extract 'b'		Extract 'r'	
	R_f value	Identity of pigment	R_f value	Identity of pigment	R_f value	Identity of pigment
pigment 1						
pigment 2						
pigment 3						
pigment 4						
pigment 5						

Table 7.2: Identification of pigments separated using R_f values.

f Table 7.3 lists the known R_f values for several photosynthetic pigments (in the solvent used), along with their colours. Use the table, and your recorded R_f values to identify each of your pigments where possible.

Pigment	R_f value in this solvent	Colour of spot
carotene	0.95	yellow
phaeophytin	0.83	yellow-grey
xanthophyll	0.71	yellow-brown
chlorophyll a	0.65	blue-green
chlorophyll b	0.45	green

Table 7.3: R_f values for pigments separated by a solvent containing 9 parts petroleum ether (80–100 °C), 1 part propanone.

g Were any pigments found in all three of the extracts?

h   Suggest why a mixture containing more pigments might be better separated over a longer distance.

i   Explain why separating the pigments over a longer distance would not affect the R_f value obtained.

j   A chromatogram can be further developed by turning it 90° and placing it into a different solvent to try to separate pigments that are very close to each other after the first solvent. Looking at your results, do you think this should be considered to further separate the pigments in your samples?

Reflection

k   Thinking about the method used, which steps could be sources of error in your results (R_f values) for this practical? How could the effect of these be reduced if you repeated the practical?

Practical 7.2: Data analysis into limiting factors for photosynthesis

Introduction

Photosynthesis is essential to life as we know it on Earth, both in the oceans and on land. Understanding how **limiting factors** affect the rate of photosynthesis is important to ensuring food webs and ecosystems remain healthy and helps explain adaptations in some producers to overcome challenges in their habitats, which can vary for the same organisms where they are very large, such as the giant kelp in Figure 7.3, which extends over a great range of depths. In this practical you will analyse data from a series of photosynthesis experiments by plotting graphs and determining when each factor is limiting on the rate of photosynthesis.

Figure 7.3: Giant kelp which is exposed to varying conditions along its length.

KEY WORD

limiting factor: the one factor, of many affecting a process, that is nearest its lowest value and hence is rate limiting; photosynthesis rate is usually limited by light intensity, temperature and/ or carbon dioxide concentration

EQUIPMENT

You will need:

- graph paper
- pencil
- ruler.

Safety considerations

There are no significant safety issues associated with this practical investigation.

> **BEFORE YOU START**
>
> a Suggest what factors may limit the rate of reactions involved in photosynthesis.
> b Suggest *two* different ways of measuring the rate of photosynthesis.
> c When investigating one variable why is it important to keep other variables constant?

> **TIP**
>
> The equation for photosynthesis is:
>
> $6CO_2 + 6H_2O \xrightarrow[chlorophyll]{light} C_6H_{12}O_6 + 6O_2$

Part 1: The effect of light intensity on the rate of photosynthesis

Method

In an experiment the amount of oxygen produced was measured when an aquatic plant was exposed to different intensities of light by varying the power supply to a lamp next to the plant. The results obtained are shown in Table 7.4.

Results

Light intensity / arbitrary units	Rate of photosynthesis / arbitrary units minute^{-1}
0	0
250	32
500	57
750	78
1000	89
1250	96
1500	100
1750	103
2000	105

Table 7.4: Results of investigation into effect of light intensity on the rate of photosynthesis.

Part 2: The effect of carbon dioxide concentration on the rate of photosynthesis

Method
In a second experiment the amount of oxygen produced was measured when an aquatic plant was placed in solutions containing different concentrations of bicarbonate which releases carbon dioxide in the solution. The results obtained are shown in Table 7.5.

Results

Carbon dioxide concentration / %	Rate of photosynthesis / arbitrary units minute^{-1}
0	22
2	68
4	91
6	110
8	110

Table 7.5: Results of investigation into effect of carbon dioxide concentration on the rate of photosynthesis.

Part 3: The effect of temperature on the rate of photosynthesis

Method
In a third experiment, the amount of oxygen produced was measured when an aquatic plant was placed in water at different temperatures. The results obtained are shown in Table 7.6

Results

Temperature / °C	Rate of photosynthesis / arbitrary units minute^{-1}
5	9
10	63
15	88
20	99
25	108

continued

7 Energy

Temperature / °C	Rate of photosynthesis / arbitrary units minute^{-1}
continued	
30	111
35	103
40	71
45	19
50	0

Table 7.6: Results of investigation into effect of temperature on the rate of photosynthesis.

Part 4: The effect of wavelength on the rate of photosynthesis

Method

In the final experiment the wavelength of light exposed to an aquatic plant was varied. The results obtained are shown in Table 7.7.

Results

Wavelength of light / nm	Rate of photosynthesis / arbitrary units minute^{-1}
380 (violet)	3
420 (violet-blue)	41
460 (blue)	19
500 (green)	16
540 (green)	7
580 (yellow-green)	17
620 (yellow-orange)	39
660 (orange)	35
700 (red)	1

Table 7.7: Results of investigation into effect of temperature on the rate of photosynthesis.

Evaluation and conclusions

d Plot a suitable graph for each of the four factors investigated.
e Describe the trend or pattern for each of the graphs you have drawn.
f Use your trends from question **e** to write a conclusion about the effect of each variable on the rate of reaction.
g From the results in Part 4 suggest why many producers are green.
h On each graph identify the point at which that factor is *not* limiting the rate of reaction any longer.
i Use all the results to suggest the optimum conditions for photosynthesis for the aquatic plant used in the experiments. How could you test this?

> **TIP**
>
> When a variable acts as a limiting factor, increases (or decreases) will change the rate of reaction. When further changes no longer change the rate of reaction (that is, the rate levels off or even drops), then a different factor has become the limiting factor.

Reflection

j To what extent do you think the data is good enough to be confident in your conclusions?

k How have these results suggested that primary productivity in the oceans may change as the levels of atmospheric carbon dioxide increase and global temperatures also increase? What other factors could be taken into consideration to answer this question better?

Practical 7.3: Gas exchange in an aquatic producer

Introduction

The overall equations for photosynthesis and aerobic respiration are the reverse of each other:

$$6\,CO_2 + 6\,H_2O \xrightarrow{photosynthesis} C_6H_{12}O_6 + 6O_2$$

$$C_6H_{12}O_6 + 6O_2 \xrightarrow{respiration} 6\,CO_2 + 6\,H_2O$$

Photosynthesis requires light, while respiration does not require light and occurs when conditions are both light and dark. In this practical you will first investigate how the presence or absence of light affect gas exchange in an aquatic plant. Then you will use your findings to adapt this method to investigate how different intensities of light affect gas exchange in an aquatic plant.

EQUIPMENT

You will need:

- 3 × boiling tubes with rubber bungs
- hydrogencarbonate indicator (approx. 60 cm³)
- 2 × pieces of pondweed (*Elodea* or *Cabomba*)
- bench lamp
- aluminium foil
- boiling tube rack (or beaker to hold the boiling tubes upright)
- measuring cylinder (25 cm³ or 50 cm³).

Safety considerations

Hydrogencarbonate indicator is an irritant. Wash any splashes on the skin with water. Wear eye protection to avoid any indicator getting in the eyes.

7 Energy

> **BEFORE YOU START**
>
> Hydrogencarbonate is a mixture of two indicators that gives a range of colours in different pHs. This is similar to how universal indicator works, but it is more sensitive to small pH changes (Figure 7.4).
>
> | increasing CO_2 in indicator | | | atmospheric CO_2 level | | | decreasing CO_2 in indicator | | |
> |---|---|---|---|---|---|---|---|---|
> | yellow | | orange | | red | | magenta | | purple |
> | pH 7.6 | pH 7.8 | pH 8.0 | pH 8.2 | pH 8.4 | pH 8.6 | pH 8.8 | pH 9.0 | pH 9.2 |
>
> **Figure 7.4:** pH ranges and colour changes for hydrogencarbonate indicator linked to changes in CO2 concentration.
>
> a Use the information in Figure 7.4 to predict what will happen to the indicator when photosynthesis occurs more than respiration.
>
> b Use the information in Figure 7.4 to predict what will happen to the indicator when respiration occurs more than photosynthesis.
>
> c One of the tests will not use any pondweed. This is a control. What is the purpose of carrying out a control experiment?

Method

1. Cut two similar lengths of pondweed and place each one into a separate boiling tube. Label the tubes 'light' and 'dark'.
2. Label the third boiling tube without any pondweed 'control'.
3. Use a measuring cylinder to add 20 cm³ hydrogencarbonate indicator to each boiling tube. Seal each boiling tube with a bung.
4. Totally cover the boiling tube labelled 'dark' with aluminium foil, to prevent any light entering the boiling tube.
5. Place all three boiling tubes into the rack or large beaker.
6. Place the bench lamp approximately 30 cm away from the boiling tubes.
7. After approximately 1 hour remove the foil and observe the colour of the hydrogencarbonate indicator in each boiling tube.

Results

d Copy and complete a table like Table 7.8.

Conditions	Colour of hydrogencarbonate indicator
light	
dark	
control	

Table 7.8: Colour changes for hydrogencarbonate indicator in different conditions.

Evaluation and conclusions

e   Look back at your predictions in 'Before you start **a** and **b**'. Were your predictions correct?

f   Explain your results.

g   Suggest why it is important that the boiling tubes are sealed with rubber bungs.

h   Suggest and explain what results you would obtain if you repeated the experiment with a small invertebrate (for example, a snail) in a test tube.

Reflection

i   How has undertaking this investigation made you realise that adaptations would be needed to vary the light intensity and obtain a wider range of results? Describe how you would vary the light intensity between 'light' and 'dark' to give at least three intermediate results.

Chapter 8
Fisheries for the future

CHAPTER OUTLINE

In this chapter you will complete investigations on:

- 8.1 Determining of size of reproductive maturity to inform minimum catch size
- 8.2 Effect of temperature on growth of whelk
- 8.3 Planning an investigation into the effect of feeding rates on the growth rates of salmon

Practical 8.1: Determining of size of reproductive maturity to inform minimum catch size

Introduction

The common whelk (*Buccinum undatum*) is a neo-gastropod mollusc found in the subtidal waters of the North Atlantic. Whelks are opportunistic scavengers that feed mainly on carrion and detect feeding opportunities with a very acute chemo-sensory system. This allows whelks to be exploited by commercial fisheries, which use baited traps. Since the early 21st century, the **fishery** for this species has undergone significant economic and geographic expansion in response to emerging markets in far-east Asia, where they are a prized food product. Regionally, the Irish Sea in north-western Europe has seen an estimated 227% increase in the total landed weight of whelk in the period 2011–2016 (Figure 8.1).

KEY WORD

fishery: the location where an aquatic species is caught

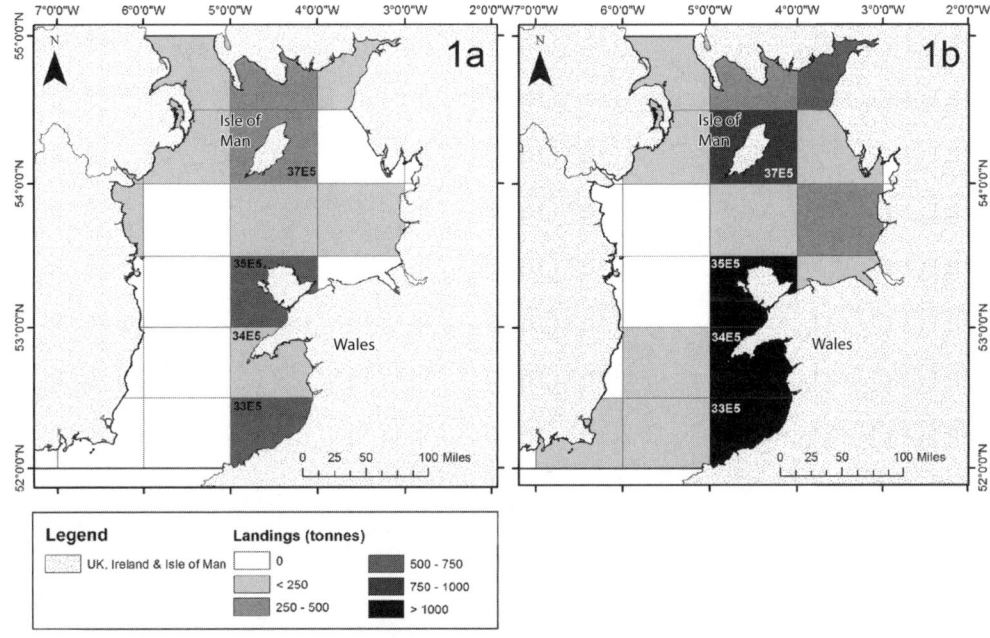

Figure 8.1: Landings of whelk (*Buccinum undatum*) in the Irish Sea in 2011 (a) and 2016 (b).

In order to avoid **overfishing** in this area, your task is to examine the reproductive biology of whelks harvested by fishermen in the Isle of Man and Wales, using an example set of genuine data which has been collected by research scientists for this purpose. You will then use this data to provide management advice to the Isle of Man and Welsh **fisheries management** departments.

EQUIPMENT

You will need:

- graph paper.

BEFORE YOU START

a What information is needed to determine suitable policies to ensure **sustainable fishing**?

b What variables may affect the size at which whelks mature?

c Suggest how scientists might determine that a whelk is sexually mature.

KEY WORDS

overfishing: rate of fishing resulting in long-term reduction in population of the species

fisheries management: protecting a fishery to enable sustainable, long-term exploitation

sustainable fishing: fishing up to the maximum sustainable yield so that future fish stocks are not at risk of being depleted

non-target species: other species caught while fishing for a particular species (this term can also be referred to as 'by-catch')

Method

The research was carried out at several sites in the Irish Sea, with the assistance of nine fishers registered in Wales, England and the Isle of Man. Each fisher followed the following procedure once a month for twelve months.

1 Two identical pots are marked and baited with similar bait.

2 The pots are lowered to the seabed on a rope with between 20–50 other pots. Record the location of the pots (latitude and longitude).

3 Leave for 24–48 hours.

4 Recover the pots from the seabed and freeze the entire contents (including under-size whelks and **non-target species**) to pass on to the scientist for analysis.

Once the contents are back at the lab the whelks are analysed.

5 All individuals are sexed (male or female), weighed (total wet weight) and measured (total shell length, TSL).

6 A random selection of individuals is inspected further for signs of sexual maturity. Just prior to the whelk spawning season, whelks show clear signs of ovary development if they are sexually mature in the 'whorl', which is made up of digestive and ovarian glands. The ovaries turn bright yellow when they are mature and full of eggs.

Results

A sample of results from whelks collected by fisheries scientists just prior to the whelk spawning season is shown in Table 8.1.

Isle of Man	Wales
68 mm ✗, 70 mm ✗, 71 mm ✗, 67 mm ✗, 85 mm ✓, 94 mm ✓, 79 mm ✓, 73 mm ✓, 64 mm ✗, 99 mm ✓, 82 mm ✗, 73 mm ✗, 86 mm ✓, 91 mm ✓, 78 mm ✓, 98 mm ✓, 89 mm ✓, 78 mm ✓, 76 mm ✗, 85 mm ✓	70 mm ✓, 69 mm ✗, 85 mm ✓, 52 mm ✗, 71 mm ✓, 73 mm ✗, 76 mm ✓, 64 mm ✓, 69 mm ✓, 69 mm ✓, 65 mm ✗, 75 mm ✓, 76 mm ✓, 84 mm ✓, 54 mm ✗, 73 mm ✓, 71 mm ✓, 80 mm ✓, 64 mm ✗, 63 mm ✗

✓ = mature, ✗ = not mature

Table 8.1: Raw data comparing size of whelks and sexual maturity off the Isle of Man and Wales.

d Draw a table to record the results in from Table 8.1, including relevant units in your headings. Include spaces to record the mean size, standard deviation and standard error for the two locations.

Evaluation and conclusions

e Calculate the mean size (total shell length) for the samples collected in Wales and the Isle of Man.

f Calculate the standard deviation (SD) of size data for whelks in the Isle of Man and Wales, using the following formula and add it to your results table.

$$SD = \sqrt{\frac{\Sigma(x - \bar{x})^2}{n - 1}}$$

in which:

SD = standard deviation

x = each individual measurement

$\bar{x}$ = the mean

n = the number of measurements made.

g Complete your results table by calculating the standard error (SE) of size data for each country, using this formula.

$$SE = \frac{SD}{\sqrt{n}}$$

h Plot a bar chart to show the mean TSL +/− 2 × SE for the two samples.

i What does your bar chart suggest about the size of whelks around the Isle of Man and Wales?

> **TIP**
>
> If the error bars on your bar chart for the two countries overlap, this indicates that any difference in the means of the two samples is not significant.

> **TIP**
>
> Total the number within each size band for the top two rows, and then cross out all the whelks that are *not* mature. This will make it easier to complete the tally for only the mature whelks.
>
> Then calculate the percentage of whelks within each size band that are mature:
>
> $$\text{percentage mature} = \frac{\text{number mature in size band}}{\text{total in size band}} \times 100$$

j   The whelks sampled can be grouped into different size ranges and the percentage in each size group with signs of reproductive maturity calculated. Copy and complete Table 8.2 using the information in Table 8.1.

	TSL / mm					
	50–59	60–69	70–79	80–89	90–99	100–109
Isle of Man: number of whelks	Tally: Number:	Tally: Number:	Tally: Number:	Tally: Number:	Tally: Number:	Tally: Number:
Wales: number of whelks	Tally: Number:	Tally: Number:	Tally: Number:	Tally: Number:	Tally: Number:	Tally: Number:
Isle of Man: mature whelks	Tally: %	Tally: %	Tally: %	Tally: %	Tally: %	Tally: %
Wales: mature whelks	Tally: %	Tally: %	Tally: %	Tally: %	Tally: %	Tally: %

Table 8.2: Collating the data into growth ranges.

> **KEY WORDS**
>
> **minimum landing size:** the smallest size permitted to be retained and sold
>
> **benthic:** on, near to or in the sediments at the bottom of a body of water

k   Draw a line graph of TSL against Percentage whelks mature (from Table 8.2), plotting the data for both Wales and the Isle of Man on the same graph. Ensure you clearly indicate which data corresponds to each site.

l   Use your graph to estimate the size at maturity for half the population in both Wales and the Isle of Man. The current **minimum landing sizes** are 45 mm (Wales) and 70 mm (Isle of Man). Are these appropriate? If these are not suitable, suggest more appropriate minimum landing sizes for each of these locations.

m   Whelks are a **benthic** community species living with many other species that are subject to benthic trawling, such as scallops. Suggest how benthic trawling could have both a negative and positive impact on whelk populations. How could the negative impacts be reduced?

> **TIP**
>
> On the x-axis, draw a scale from 50 to 110 and plot the points at the mid-point for each set of data; that is, for 50–59 plot at 55, for 60–69 plot at 65, etc.

Reflection

n   Thinking about the time it took you to analyse the data provided and the range of data you analysed, how could you adapt the calculations to handle a larger set of data, to increase the reliability of the results and the accuracy of the calculations?

Practical 8.2: Effect of temperature on growth of whelk

Introduction

Accurate scientific data is important to ensure that appropriate policies are set to manage fisheries in a sustainable way. The growth rates of many species, including whelks, are affected by many variables that can lead to significant variation in growth rate even

8 Fisheries for the future

> **BEFORE YOU START**
>
> a Why does temperature affect the growth rate of many marine organisms?
> b What other variables may affect the growth rate of marine organisms?
> c Suggest how changes over a full year could affect the growth rate of organisms, and why this may lead to the development of growth rings.
> d Why would taking seabed temperature readings on the particular day that the whelks are fished have limited importance?

between populations only a short distance apart. The impact of temperature is known to affect the growth rates of many marine species. This information can be important in ensuring that fishing management strategies are set appropriately. In this investigation, you will look at evidence that temperature affects the growth rate of whelks.

Method

The research was carried out with whelks collected at several sites in the Irish Sea with the assistance of commercial fishers. The location of each sample provided was recorded (latitude and longitude). The whelks were frozen on landing and sent to the scientists. When the whelks were back at the lab they were thawed and analysed.

1   The total shell length was recorded to the nearest 0.1 mm.

2   To determine the age of each whelk, one **statolith** was extracted and examined using a microscope to observe and count the annual growth rings (see Figure 8.2). Only whelks with clear and distinct growth rings were included in the data.

> **KEY WORD**
>
> **statolith:** a calcareous particle found in the nervous system in many molluscs, which helps them to determine the direction of gravity

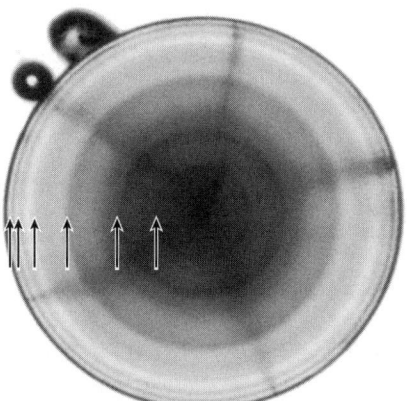

Figure 8.2: Photomicrograph showing growth rings in a whelk statolith.

3   Sea-floor temperatures were obtained for each location, taking into account many variables. These gave average sea-floor temperatures for the location for the six years prior to collecting the whelks.

4   The sea-floor temperatures were converted into total annual 'degree days' (sum of the average daily sea-floor temperature over 365 days).

Results

Isle of Man			Swansea Bay (Wales)		
Ring no.	Statolith diameter / μm	TSL / mm	Ring no.	Statolith diameter / μm	TSL / mm
0 (hatching)	86	5	0 (hatching)	69	3
1	138	15	1	125	12
2	177		2	165	23
3	236	52	3	213	42
4	265	69	4	253	63
5	282		5	270	73
6	298		6	277	78
7	305	97	7	280	80
8	310		8	280	80
9	315		9	280	80

Table 8.3: Diameters of growth rings in statoliths of whelks from the Isle of Man and South Wales.

Evaluation and conclusions

e Table 8.3 shows the diameter of each ring (in μm) for two statolith samples, one from the Isle of Man and one from Swansea Bay in Wales. Use the equation below, which estimates the total shell height (TSL in mm) based on the diameter of the ring, to complete the gaps in the table.

$$\text{estimated TSL (mm)} = \left(\frac{\text{statolith}}{43.43}\right)^{2.34}$$

f Draw a line graph to show the size (TSL) at age (Ring no.) of the two samples in Table 8.3. Draw all data on the same graph.

g From what you have found out about size at maturity (Practical 8.1) and the growth rates of whelk in your graph from above, at what age does the average female first lay eggs in each place?

    i Isle of Man

    ii Wales.

h In marine ecosystems, environmental factors can have significant effects on the biological characteristics of an animal. Among others, sea-bottom-temperature (SBT °C) is an important factor that effects how quickly, and how large whelks can grow.

Complete the information from the missing box in Table 8.4 with the information on the maximum size (L_{MAX}) of whelk that you have already calculated and presented above. Another variable 'degree-days' is given to you, and is the cumulative sum of daily temperatures recorded at the sea floor where the whelks were sampled.

Area	Degree-days	L_{MAX}
Isle of Man	4000	
Anglesey	4025	97
Nefyn	4175	87
Bardsey	4055	90
Carmarthen Bay	4185	82
Swansea	4270	80

Table 8.4: Maximum shell length of whelks and cumulative daily temperatures for six locations in the Irish Sea.

i   Plot a line graph of degree-days (*x*-axis) against L_{MAX} (*y*-axis).
j   Describe the relationship between temperature (as determined by annual degree-days) and maximum size of whelks.
k   If climate change leads to warmer seawaters, what implications does this have for whelk fisheries in the Irish sea?

Reflection

l   How has this activity helped you understand the challenges faced in managing fishing to ensure that that it is sustainable?

Practical 8.3: Planning an investigation into the effect of feeding rates on the growth rates of salmon

Introduction

Aquaculture businesses raise an aquatic species to maturity to sell for profit. To maximise profitability, the business must offset the cost of feed against any increase in growth rates. This investigation is to try to determine the most profitable level of feeding salmon so that they grow quickly in an aquaculture venture, such as in the setup shown in Figure 8.3.

You will plan an investigation that could be carried out in a laboratory in large tanks containing salmon. (You will not be carrying this out yourself, so you are not limited to equipment that is available to you.) Your aim is to find the optimum feeding rate for salmon growth.

Figure 8.3: A typical salmon farm enclosure in Norway.

> **BEFORE YOU START**
>
> a   Suggest what impacts the feeding intensity for an aquaculture business might have on the local environment.
>
> b   How could scientists establish a suitable range of feeding rates to investigate in more detail?
>
> c   Write a hypothesis that you could investigate.

> **TIP**
>
> To measure a 'rate', you need to measure both what has changed and the time over which it has changed.

Planning

d   Plan your investigation. Your plan should:
   - include the hypothesis you are investigating
   - identify all the key variables
   - include a clear method that is easy for others to follow
   - include a suitable table for recording results
   - describe how you would analyse the results collected, including a sketch of the graph you would plot with the axes labelled and a prediction for the pattern of results
   - be safe and ethical.

> **TIP**
>
> Look back at earlier planning exercises that developed your writing skills in planning an investigation for prompts on how to organise your answer.

Chapter 9
Human impacts

> **CHAPTER OUTLINE**
>
> In this chapter you will complete investigations on:
> - 9.1 Planning an investigation into marine plastics pollution
> - 9.2 Modelling the greenhouse effect
> - 9.3 Monitoring invasive species

Practical 9.1: Planning an investigation into marine plastics pollution

Introduction

Plastic waste and **pollution** have been shown to have reached virtually all parts of the oceans, from the visible waste seen on shorelines (see Figure 9.1) and floating in gyres in the middle of ocean basins, to the deepest ocean trenches. Increasing awareness of this problem, and the uncertainty of the long-term effects of this pollution on the health of marine organisms and food chains, is helping to encourage the reduction in use of single-use plastics through both government policy and voluntary actions by changing our habits as consumers.

In this exercise you will plan an investigation to help government to evaluate the effectiveness of a policy related to reducing the amount of plastic waste entering the oceans.

> **KEY WORD**
>
> **pollution:** when substances added to the environment have a harmful effect

> **BEFORE YOU START**
>
> a   What policies (such as laws or incentives) have been introduced in your region to reduce the amount of single-use plastics used, or to attempt to reduce the amount of plastic pollution reaching the oceans?
>
> b   What impacts of plastics and microplastics do these measure help to reduce?
>
> c   Choose one of the policies. What data would have to be collected to show the difference this policy is making to plastic pollution?
>
> d   Write a hypothesis that you could investigate to test the effectiveness of your chosen policy.

Figure 9.1: Clearing discarded plastic waste from a beach.

Planning

e Plan an investigation that could be used to collect evidence to show whether your chosen policy is effective. Your plan should:
- include the hypothesis you are investigating
- identify all the key variables
- include a clear method that is easy for others to follow
- include a suitable table to record results in suggest a suitable type of graph to present the data
- describe how you would analyse the results
- be safe and consider the ethical treatment of any organisms and habitats in the areas investigated.

Practical 9.2: Modelling the greenhouse effect

Introduction

This experiment shows how the concentration of carbon dioxide in the atmosphere can affect the rate of warming in the atmosphere. It should show the difference between normal levels of carbon dioxide and additional carbon dioxide by heating two different 'atmospheres' in similar beakers under similar conditions. We can observe the difference this has on the rate of warming with more carbon dioxide added to only one of the beakers and comparing the temperature rises in the two beakers.

EQUIPMENT

You will need:

- 2 × 500 cm^3 beakers
- disks of aluminium foil (painted on the upper surface with black matt paint) cut to fit the base of the beakers
- 2 × 0–110 °C thermometers (or 2 × temperature probes connected to a datalogger)
- 2 × 60 W incandescent or halogen lamps
- 250 cm^3 conical flask fitted with a delivery tube and 100 cm^3 syringe
- marble chips (approximately 10 g)
- hydrochloric acid, 2 M (100 cm^3)
- stopwatch or timer.

Safety considerations

The bulbs will get very hot in use. Take care to allow them to cool completely before moving them. Hydrochloric acid (2 M) is an irritant. Wear goggles while handling it and wash any splashes off the skin.

BEFORE YOU START

a What processes release and remove carbon dioxide into and from the atmosphere?
b What does the bulb in the experiment represent?
c What does the painted foil in the experiment represent?

Method

Before starting the experiment, read through carefully and prepare a results table to record all your results in (see question **d**).

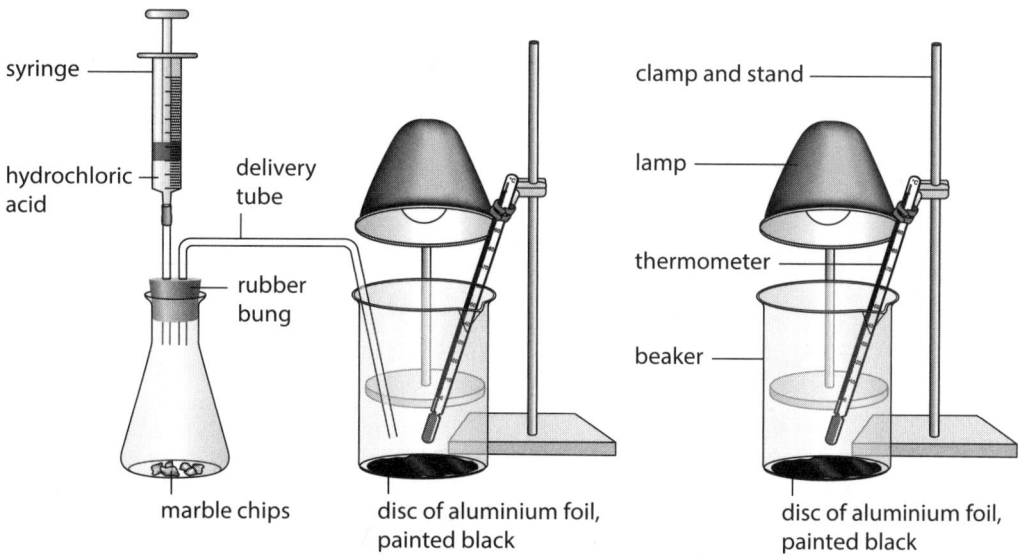

Figure 9.2: Diagram showing how to set up apparatus.

1. Set up the apparatus as shown in Figure 9.2.
    - i   Place a piece of aluminium foil that has been painted black in the bottom of each of the two beakers.
    - ii Place the beakers side by side below the bulbs so they are equally heated by the bulb.
    - iii Clamp two thermometers (or temperature probes) so they are held about 2 cm above the foil in each beaker.
2. Switch on the bulbs and adjust the position of the beakers / thermometers so both give similar temperature readings. This may take 10–15 minutes.
3. Set up the conical flask and delivery tube, as shown, with marble chips and hydrochloric acid in the syringe. (The marble chips contain calcium carbonate. This reacts with hydrochloric acid to produce bubbles of carbon dioxide.)
4. When the temperature readings in both beakers stabilise (remain fairly constant), slowly add all the acid to the marble chips to generate carbon dioxide in beaker 2. Start the stopwatch.
5. Record the temperature in both beakers every 30 seconds for 10–15 minutes, recording the temperature in each of the beakers.

Results

- **d** Design a single results table to record the results for both beakers.
- **e** Plot the results you obtain on the same graph so you can compare the results.

Evaluation and conclusions

f Which beaker contained extra carbon dioxide, and what did this do to the temperature in that beaker?

g How do the temperature changes after switching on the lamps and then after adding extra carbon dioxide explain the difference between the natural greenhouse effect and enhanced greenhouse effect? How does this lead to global warming?

h Suggest reasons why there was a delay in the temperature's changing and why the temperatures in the beakers may be different before adding the carbon dioxide.

i How could you check the reliability of your results?

j The **greenhouse effect** is an important phenomenon for life on Earth. Discuss evidence for and against the hypothesis that human activity significantly contributes to **global warming**.

k If global warming continues, what possible impacts could this have on marine environments?

Reflection

l Do you think that the evidence collected in this experiment supports the argument that humans are contributing to global warming? How could the experiment be adapted to give more evidence that human activity is to blame?

> **KEY WORDS**
>
> **greenhouse effect:** the heating of the atmosphere due to the presence of carbon dioxide and other gases; these gases prevent infrared radiation being re-emitted into space
>
> **global warming:** the observed and projected increases in the average temperature of Earth's atmosphere and oceans due to an enhanced greenhouse effect

Practical 9.3: Monitoring invasive species

Introduction

Invasive species are organisms that become present in a region where they have not previously lived. By definition, these species are different in different regions around the world. Some species may have little effect in their new ecosystem, but others can pose a significant threat to the biodiversity of the area and potentially have catastrophic effects on both ecosystems and the economic success of fisheries. In this task, you will examine signs of invasive species in your region, using identification cards or leaflets to identify non-native species. If possible, this should be along a coastline, but if you cannot get to a coast you can look for invasive species inland instead.

> **EQUIPMENT**
>
> **You will need:**
> - set of species ID cards
> - clipboard or notepad
> - hand lens
> - pencil
> - camera.

9 Human impacts

Safety considerations

Follow your teacher's instructions carefully. Shorelines can be dangerous.

> **BEFORE YOU START**
>
> **a** Describe two different causes / methods for a marine species to move to a new region.
> **b** Describe safety precautions to take while carrying out a fieldwork survey.
> **c** Why is it important that national / international surveys follow a standard method?

Method

1 Divide the species ID cards among the group. Study the photograph(s) and information about each species to ensure that you know in what conditions it is most likely to be found and how to identify it. (Take note of any key similarities to and differences from related species.)

2 Spend 20 minutes searching an area for all the species you have cards for, taking care not to tread on organisms and carefully replacing any rocks moved so not to damage any organisms.

3 Take photographs of any examples you find. Make a note of where they were found.

4 At the end of the 20 minutes, record the abundance of each species in the area. You may wish to check any photographs with your teacher, if you are unsure about identification. Record your results in a copy of Table 9.1.

Results

Species	Abundance*			
	A	F	R	NF
*KEY – For abundance of organisms use the following scale: A = ABUNDANT (This species is found widely over the area surveyed.) F = FREQUENT (This species is found in multiple locations over the area surveyed.) R = RARE (This species was seen but there were very few individuals.) NF = NOT FOUND (This species was not identified at all in the area surveyed.)				

Table 9.1: Results table for Practical 9.3.

Evaluation and conclusions

d   If survey results are available for previous years, compare your data with those results. Describe any changes in the results and what they suggest.

e   What are the limitations of the data collected?

f   How could the quality of the data collected be improved?

g   Why is it important to monitor whether invasive species are present in an ecosystem?

h   Describe how information from surveys like this can help governments and conservation organisations.

i   What other organisations / stakeholders can help to reduce the risk of introducing invasive species, and how could they do this?

Reflection

j   How could the method for the survey be improved to make results more consistent each year?

> Glossary

Command words

calculate: work out from given facts, figures or information

comment: give an informed opinion

compare: identify / comment on similarities and/or differences

contrast: identify / comment on differences

define: give a precise meaning

describe: state the points of a topic / give characteristics and main features

discuss: write about issue(s) or topic(s) in depth in a structured way

evaluate: judge or calculate the quality, importance, amount or value of something

explain: set out purposes or reasons / make the relationships between things clear / say why and/or how and support with relevant evidence

give: produce an answer from a given source or recall/memory

identify: name / select / recognise

justify: support a case with evidence / argument

outline: set out the main points

predict: suggest what may happen based on available information

state: express in clear terms

suggest: apply knowledge and understanding to situations where there are a range of valid responses in order to make proposals / put forward considerations

Key words

abiotic: non-living components of the ecosystem where chemicals are inorganic

abiotic factors: the environment's geological, physical and chemical features; the non-living part of an ecosystem

aerobic respiration: the release of energy from glucose or another organic substrate in the presence of oxygen; the waste products are carbon dioxide and water

anomaly: a result or observation that deviates from what is normal or expected; in experimental results, it normally refers to one repeated result that does not fit the pattern of the others

aquaculture: the rearing of aquatic animals and plants for human consumption or use

artificial reef: an underwater structure built by humans to mimic the characteristics of a natural reef

atomic mass: the mass of an atom that is approximately equal to the number of protons and the number of neutrons added together

atomic number: the number of protons contained in the nucleus of an atom

benthic: on, near to or in the sediments at the bottom of a body of water

benthic floor: the habitat at the bottom of the ocean

benthic trawling: a fishing method that drags a net along the seabed; wooden boards at the front of the net keep the net open and stir up the seabed, causing damage

binomial nomenclature: the two-part Latin name given to each species comprising the genus followed by the species

biodiversity: a measure of the species, genetic and ecosystem diversity of different species

biological drawing: a scientific drawing that records an image and important features of the specimen

biomass: the mass of living material in an area; it can be measured as dry mass (without the water) or wet mass (with the water)

biotic: living components of the ecosystem where chemicals are organic

biotic factors: the living parts of an ecosystem, which includes the organisms and their effects on each other

blue carbon: carbon stored in marine ecosystems

Calvin cycle: the series of reactions that occur during the light independent stage of photosynthesis; it converts carbon dioxide and other substances into glucose

carbon cycle: the range of processes that involve the chemical and physical changes to carbon resulting in carbon transforming through a range of substances, including in the atmosphere, living organisms and rocks

carbonate rock: a rock with a major component of minerals containing carbonate ions (for example, limestone contains mostly calcium carbonate)

categoric variable: a variable that is not continuous and has a value that is a name or label such as the colours red, blue and green

cell surface membrane: a biological membrane that separates the internal contents of a cell from its external environment

cell wall: a layer that surrounds some types of cells and gives strength and support; plant cell walls are made of cellulose

chi-squared (χ^2) test: a statistical test to measure how expected outcomes compare to actual outcomes

chloroplast: the photosynthetic organelle in eukaryotes

chromatography: a technique used to separate substances by their solubility

circulatory system: an organ system comprising the heart, blood vessels and the blood; its role is to carry useful substances around the body to all cells and move waste products away from the cells to be removed

community: all the different populations interacting in one habitat at the same time

complex life cycle: cycle in which the pre-reproductive and reproductive stages are very different and may change morphology, habitat, and diet

continental crust: the thicker, less dense crust that makes up the foundation of the continents

continuous belt transect: use of a quadrat to collect population information along a transect without any gaps

continuous variable: one which can take any value (for example, temperature, time, concentration)

control variables: variables that are not being tested but that must be kept the same in case they affect the experiment

coral bleaching: whitening of coral that results from the loss of a coral's symbiotic zooxanthellae

correlation: the tendency of two variables to change together in either the same or opposite directions

covalent bond: chemical bond that involves the sharing of electron pairs between atoms

density: a measure of the mass of a defined volume of water

dependent variable: the variable being measured in an experiment

desalination plants: industrial facilities that remove salts from seawater to manufacture fresh water

dichotomous key: an identification tool utilising a series of choices between alternative characters, with a direction to another stage in the key, until the species is identified

diffusion: the random movement of particles (or molecules) from a higher concentration to a lower concentration (down a concentration gradient); it is a passive process, not requiring the input of energy

distribution: the variation of a population across an area

divergent boundary: where two tectonic plates are moving away from each other

erosion: a natural process where material is worn away from the Earth's surface and transported elsewhere

error bars: lines through points on graphs to represent the uncertainty of a measurement

ethical issue: when an individual or group has to choose between alternatives that can be evaluated as either right or wrong

euryhaline: organisms that can tolerate wide changes in the salinity of water

facultative halophytes: plants that can grow in saline and non-saline habitats

fisheries management: protecting a fishery to enable sustainable, long-term exploitation

fishery: the location where an aquatic species is caught

food chain: a way to describe the feeding relationships between organisms

food web: a way to show all the different feeding relationships in an ecosystem

fossil fuels: buried organic materials from dead plants and animals which have been converted into oil, coal or natural gas by exposure to heat and pressure in the Earth's crust

gas (gaseous) exchange: the uptake of oxygen and release of carbon dioxide by cells or other surfaces

gills: the gaseous exchange surfaces of fish

global ocean conveyor belt: constantly moving systems of deep-ocean water driven by thermohaline circulation

global warming: the observed and projected increases in the average temperature of Earth's atmosphere and oceans due to an enhanced greenhouse effect

Golgi body: cell organelle that modifies proteins

greenhouse effect: the heating of the atmosphere due to the presence of carbon dioxide and other gases; these gases prevent infrared radiation being re-emitted into space

habitat: the natural environment where an organism lives

halocline: a layer of water below the mixed surface layer where a rapid change in salinity can be measured as depth increases

hydrogen bond: a weak bond between two molecules due to the electrostatic attraction between a hydrogen atom in one molecule and an atom of oxygen, nitrogen or fluorine in the other molecule

independent variable: the variable being changed in an experiment

interrupted belt transect: use of a quadrat to collect population information along a transect with regular gaps between samples

invasive species: species that have become established in an area that is not their normal habitat due to human activity

ionic bond: chemical bond that involves the attraction between two oppositely charged ions

kite graph: a graph of the distribution and abundance of organisms in the littoral zone that allows zonation patterns to be easily seen

***K*-strategy:** producing few offspring but providing a large amount of parental investment

lamellae: addition branches of gill filaments that increase the surface area of the gills; they are sometimes called secondary lamellae

large permanent vacuole: a membrane bound organelle that is present in all plant and fungal cells; it contains cell sap

light dependent stage: the stage in photosynthesis whereby light energy is harvested; occurs in the thylakoid membranes of chloroplasts and produces ATP and reduced NADP

light independent stage: the stage in photosynthesis whereby carbon dioxide is converted into glucose by the Calvin cycle; occurs in the stroma of chloroplasts

limiting factor: the one factor, of many affecting a process, that is nearest its lowest value and hence is rate limiting; photosynthesis rate is usually limited by light intensity, temperature and/or carbon dioxide concentration

Lincoln index: a mathematical equation that can use the mark–release–recapture data to estimate the population size

marine protected area (MPA): an area of ocean or coastline where restrictions have been placed on activities; the levels of restriction may vary, some may be no-take areas where no fishing is permitted, others may allow some fishing, some may ban all access to unauthorised people, while others may allow restricted access

marine toxins: poisonous chemicals that can contaminate seawater

mark–release–recapture: a method to estimate the population size of mobile species

mean: the sum of a number of items of data, divided by the total number of items of data added.

meniscus: the upward or downward curve at the surface of a liquid where it meets a container

microplastics: plastic particles that are less than 5 mm in diameter

mid-ocean ridge: a mountain range with a central valley on an ocean floor at the boundary between two diverging tectonic plates, where new crust forms from upwelling magma

minimum landing size: the smallest size permitted to be retained and sold

mitochondrion: (plural: mitochondria) the organelle in eukaryotes in which aerobic respiration takes place

monomer: the smallest unit of a polymer; monomers are able to join chemically to form longer molecules

motile species: an organism that can move around in its habitat and is not fixed in place

neap tide: a tide that occurs when the Moon and Sun are at right angles from each other, causing the smallest tidal range

niche: the role of a species within an ecosystem

non-target species: other species caught while fishing for a particular species (this term can also be referred to as 'by-catch')

nucleus (cell): membrane bound organelle that contains the genetic material of a cell

null hypothesis (H_0): there is no correlation between the two sets of variables

obligate halophytes: plants that can only survive in a saline environment

oceanic crust: the dense, basaltic layer of crust that makes up the bottom of the ocean basins

osmoconformer: organisms that have an internal body fluid salinity that it is the same salinity as external water

osmoregulator: organisms that regulate the internal salinity of their body fluids within a narrow range

osmosis: the movement of water from a higher water potential to a lower water potential across a selectively permeable membrane

overfishing: rate of fishing resulting in long-term reduction in population of the species

permeability: how well water flows through a substrate

pH: a figure expressing the acidity or alkalinity of a solution on a logarithmic scale

photosynthesis: the process of using light energy to synthesise glucose from carbon dioxide and water to produce chemical energy

photosynthetic pigments: pigments such as chlorophyll that are used to absorb light during photosynthesis

pollution: when substances added to the environment have a harmful effect

polymer: a large molecule made from many repeating sub-units

pooter: a bottle for collecting small invertebrates, having one tube through which they are sucked into the bottle and another, protected by muslin or gauze, which is sucked

population: all the individuals of the same species that live at the same place and time

porous: substrate with holes that allow for the passage of air and water

primary productivity: the rate of production of new biomass through photosynthesis or chemosynthesis

pyramid of biomass: a diagram that shows the biomass present in each trophic level of a food chain

pyramid of energy: a diagram that shows the amount of energy in each trophic level of a food chain

pyramid of numbers: a diagram that shows the number of organisms in each trophic level of a food chain

quadrat: a square to mark an area, often divided into smaller squares; can be different sizes such as 1 metre × 1 metre or 0.5 metre × 0.5 metre

random sampling: samples are taken at random places within the sample site

range: the maximum variation of a data set from the lowest to the highest value

ribosomes: small organelles that are involved in the synthesis of proteins

rough endoplasmic reticulum (rER): a network of flattened membranous sacs, covered with ribosomes, that runs through the cytoplasm of a cell; proteins are synthesised in it before being transported to the Golgi body in vesicles

r-strategy: providing large numbers of offspring while providing little parental investment

salinity: a measure of the quantity of dissolved solids in ocean water, represented by parts per thousand (ppt) or ‰

sedimentary rock: rock formed by the deposition of particles on the ocean floor

sedimentation rate: the rate at which sediment is deposited on the sea floor

sewage: liquid and solid waste material such as wastewater or urine

simple life cycle: cycle in which pre-reproductive and reproductive stages are very similar to one another

Simpson's index of diversity (D): a biodiversity measure that accounts for both species richness and eveness

smooth endoplasmic reticulum (sER): a network of flattened membranous sacs that is found in the cytoplasm of cells; it is distinguished from the rough endoplasmic reticulum by its lack of ribosomes; its main function is the synthesis of lipids

solubility: the ability of a solute to dissolve within a solvent (such as water)

solute: a solid that dissolves in a solvent

SONAR: a method that is used to detecting underwater objects by the reflection of sound waves

Spearman's rank correlation (r_s): a mathematical tool used to find out if there is a correlation between two sets of variables, when they are not normally distributed

species: a group of similar organisms that can interbreed naturally to produce fertile offspring

spreading rate: the rate at which the new seafloor moves apart at a mid-ocean ridge

spring tide: a tide that occurs when the Sun and Moon are aligned, causing the largest tidal range

standard deviation: a measure of the spread of data about the mean value

standard error: a measure of how much a sample mean deviates from the true mean

standardised variable: a variable that is kept constant during an experiment

Glossary

statolith: a calcareous particle found in the nervous system in many molluscs, which helps them to determine the direction of gravity

stenohaline: organisms that cannot tolerate wide changes in the salinity of water

substrate: the material that makes up the sea floor, such as rocks, sand, silt, etc.

sustainable fishing: fishing up to the maximum sustainable yield so that future fish stocks are not at risk of being depleted

systematic sampling: samples are taken at fixed intervals along the transect

taxonomic hierarchy: the classification of the species within living organisms by describing the domain, kingdom, phylum, class, order, family, genus and species

thermocline: a layer between two layers of water with different temperatures

transect: a rope or tape marked at regular intervals that sets standard distances for study of the distribution of marine organisms

trophic efficiency: the efficiency with which energy is transferred from one trophic level to the next

trophic level: the position an organism occupies in the food chain or food web

trophic level transfer efficiency (TLTE): measures the amount of energy that is transferred between trophic levels

turbidity: the level of transparency loss water has due to the presence of suspended particles in the water; the higher the turbidity, the harder it is to see through the water

unstable habitat: a habitat that has moving or shifting substrate making it difficult to attach to

ventilation: the exchange of gases (oxygen and carbon dioxide) into and out of organisms; the term also describes a method of achieving this (for example, breathing in mammals and ram or pumped ventilation in fish)

viviparous reproduction: [plants] a reproductive strategy where the seed develops into a young plant while still attached to the parent plant

water potential: the potential energy of water in a solution compared to pure water; water will move by osmosis from a higher water potential to a lower water potential

weathering: the wearing down or breaking of rocks through physical, chemical or organic means

Wentworth scale: a way to describe different types of sand based on their grain size

World Ocean: the combination of all major oceans into one large, interconnected body of water that encircles the world's continents

zooxanthellae: symbiotic, photosynthetic dinoflagellates living within the tissues of many invertebrates

95% confidence limits: range of values between which there is 95% confidence that the mean lies

> Acknowledgements

The authors and publishers acknowledge the following sources of copyright material and are grateful for the permissions granted. While every effort has been made, it has not always been possible to identify the sources of all the material used, or to trace all copyright holders. If any omissions are brought to our notice, we will be happy to include the appropriate acknowledgements on reprinting.

Thanks to the following for permission to reproduce images:

Cover Georgnroll/Getty Images

Section 1: Steve Christo Corbis/GI; Formiktopus/GI; Avalon/GI; Chicago Tribune/GI; Carol Yepes/GI; DR Keith Wheeler/Science Photo Library; Edward Kinsman/Science Photo Library; **Section 2:** George Rose/GI; GIPhotoStock/GI; Jose A. Bernat Bacete/GI; Sandra Standbridge/GI; Nature Photographers Ltd/Alamy Stock Photo; Tetra Images/GI; Ullstein bild/GI; Figure 1. Landings of whelk (Buccinum undatum) in the Irish Sea in 2011 (1a) and 2016 (1b). Data Source: iFISH2 Database. Reprinted from Fisheries Research, 204, Emmerson et al., The complexities and challenges of conserving common whelk (Buccinum undatum, L.) fishery resources: Spatio-temporal study of variable population demographics within an environmental context, 125-136, 2018, with permission from Elsevier; Irish Sea Buccinum undatum (common whelk) case study data and photos were provided by the Fisheries & Conservation Science Group, School of Ocean Sciences, Bangor University, Wales, courtesy of Jack Emmerson; NurPhoto/GI; Jack Guez/GI.

Key: GI= Getty Images

We would like to thank Mary-Jane Watson, science technician at University College Isle of Man, for kindly helping to test practical activities and prepare resources for students to trial.

We would also like to thank AS and A Level Marine Science students at University College Isle of Man for their generous help testing some of the practical activities in this workbook. The students include Jake Christian, Jacob Duggan, Molly Cracknell, Jonathan Ireland, Erin Lacy, Russel Miguel, Jamimah Nicoll, Kristian Orchard, Danny Peel, Elan Shipton, George Thwaites and Brandon Wade.

In addition, we would like to thank Dr Lara Howe, Marine Officer at the Manx Wildlife Trust for her contributions to the development of Practical Activity 9.3.

The case study data and photographs in Practical activities 8.1 and 8.2 were provided by the Fisheries & Conservation Science Group, School of Ocean Sciences, Bangor University, Wales, courtesy of Jack Emmerson. This work was funded by the Isle of Man Government's Department of Environment, Food and Agriculture